Applied Logistic Regression

DAVID W. HOSMER, Jr.
Professor of Biostatistics
School of Public Health and Health Sciences
University of Massachusetts
Amherst, Massachusetts

STANLEY LEMESHOW
Professor of Biostatistics
School of Public Health and Health Sciences
University of Massachusetts
Amherst, Massachusetts

WILEY

A WILEY-INTERSCIENCE PUBLICATION

JOHN WILEY & SONS
New York   Chichester   Brisbane   Toronto   Singapore

Library of Congress Cataloging in Publication Data:
Hosmer, David W.

    Applied logistic regression/David W. Hosmer, Jr., Stanley
Lemeshow.

       p. cm.–(Wiley series in probability and mathematical
statistics. Applied probability and statistics section)

    "A Wiley-Interscience publication"

    Bibliography : p.

    ISBN 0-471-61553-6

    1. Regression analysis. I. Lemeshow, Stanley. II. Title.
III. Series.

QA278 2.H67 1989

519.5'36–dc20

                                      89-31893
CIP

Printed in the United States of America

10 9 8 7 6 5 4 3 2 1

To Our

Wives: Trina Hosmer   –   Elaine Lemeshow

Children: Wylie and Trina Hosmer   –   Jenny, Adina, and Steven Lemeshow

PREFACE

The logistic regression model has been in use in statistical analyses for many years; but it was not until Truett, Cornfield, and Kannel (1967) used the model to provide a multivariate analysis of the Framingham heart study data that its full power and applicability were appreciated. Since that landmark paper the logistic regression model has become the standard method for regression analysis of dichotomous data in many fields, especially in the health sciences. Nearly every issue of such major journals as *The American Journal of Epidemiology, The American Journal of Public Health, The International Journal of Epidemiology*, and *The Journal of Chronic Diseases* has articles whose analyses are based on the logistic regression model.

In reviewing papers in these journals as well as many others, it became clear that the quality of the use of the method is not on the same level as analyses using linear regression. While the inclusion of logistic regression routines in major computer software packages such as GLIM, BMDP, SAS, SPSS-X, and others has allowed widespread use of the method, it seems that many users do not understand that they are, in fact, doing regression. Thus, the reporting of results is highly variable and assessment of model adequacy is, for the most part, nonexistent. This could, in part, be due to the lack of an introductory text on the model.

The literature on logistic regression is large and growing rapidly. A few textbooks that cover aspects of logistic regression include Breslow and Day (1980), Cox (1970), Kleinbaum, Kupper, and Morgenstern (1982), and Schlesselman (1982). In each of these texts logistic regression is not the central focus. In addition, many of the techniques for application of the method and interpretation of the results may only be found in statistical literature that is beyond the comprehension of many potential users.

The primary objective of this textbook is a focused introduction to the logistic regression model and its use in methods for modeling the relationship between a dichotomous outcome variable and a set of covariates. The intended audience includes beginning graduate students in biostatistics, graduate students in epidemiology, and working professionals who wish to learn how to model a binary outcome variable. We view the book as a textbook for a one-quarter or a semester course on the model and its application. We assume that students have a solid foundation in linear regression methodology and contingency table analysis through Mantel–Haenszel methods.

The approach we will take is to develop the model from a regression analysis point of view. This is accomplished by approaching logistic regression in a manner analogous to what would be considered good statistical practice for linear regression. This differs from the approach used by other authors who have begun their discussion from a contingency table point of view. While the contingency table approach may facilitate the interpretation of the results, we believe that it obscures the regression aspects of the analysis. Thus, discussion of the interpretation of the model is deferred until the approach to the analysis is firmly established.

To analyze the data sets used in this book we have used a number of software packages having logistic regression capabilities. These include BMDP [Dixon, 1987], EGRET [Statistics and Epidemiology Research Corp., 1988], GLIM [Numerical Algorithms Group, 1987], SAS [SAS Institute Inc., 1988], and SYSTAT [Wilkinson, 1987]. In general, the results produced were the same regardless of which package was used. We did note some differences, though not important ones, in the values of estimated standard errors of estimated logistic regression coefficients. Reported numeric results have been rounded from figures obtained from computer output and thus may differ slightly from those that would be obtained in a replication of our analyses or from calculations based on the reported results. To avoid either knowingly or unknowingly supporting any particular package, the program used to obtain results will not be specified. When features or capabilities of the programs differ in an important way, we will note them by the names given rather than by their bibliographic citation.

The text includes several data sets which are the source of the examples and exercises.

The manuscript was prepared for camera ready production on an Apple Macintosh™ and an Apple LaserWriter Printer. The basic word processing program used was Microsoft® Word (Version 3.1) [Microsoft Corporation (1987)]; all formulae were produced using MathWriter™ (Version 1.4) [Cooke and Sobel (1986)]; graphs were produced using the SYSTAT Graph Module (Version 3.2) [Wilkinson (1987)]; and tables were produced in Microsoft® Excel (Version 1.5) [Microsoft Corporation (1987)]. MacDraw II (Version 1.1) [CLARIS Corporation (1987)] was indispensable for adjusting the spacing in formulae and the general appearance of graphs and tables.

We are appreciative of the efforts of the many individuals who assisted us in the preparation of this book. In particular, Bev Decker has worked long hours and has had to draw on her considerable skills to carry the major word processing responsibilities for this project. These skills were invaluable due to our decision to prepare the manuscript in camera ready format. Her discovery of techniques not explained in the manuals became the rule rather than the exception.

We also appreciate the efforts of all those who carefully read earlier versions of our manuscript. This includes two of our graduate students: Bob Harris was careful in his reading and extremely attentive to detail; and Gordon Fitzgerald made extremely helpful comments throughout the text. Our colleague, Ed Stanek, made suggestions which also improved the book. We would especially like to express our appreciation of the efforts of Sander Greenland for his careful reading, invaluable suggestions, and numerous discussions of specific details and problems. The book has benefited greatly from his contributions.

There have been many other contributors to this book. Data sets were made available by our colleagues, Harris Pastides, and Jane Zapka, as well as by

Doctors Daniel Teres and Laurence E. Lundy at Baystate Medical Center in Springfield, Massachusetts. Professor Tor Bjerkedal and the staff of the Institute of Preventive Medicine of the University of Oslo and Mr. Petter Laake, Section on Medical Statistics, also of the University of Oslo were very helpful when one of us (DWH) visited during the winter of 1988 when a major part of the book was written. Comments of our many students and colleagues, both at the University of Massachusetts as well as the New England Epidemiology Summer Program, have been invaluable.

Finally, we would like to thank Beatrice Shube, Kate Roach, and the production staff at John Wiley and Sons for their help in bringing the project to completion.

DAVID W. HOSMER, JR.
STANLEY LEMESHOW

Amherst, Massachusetts
April, 1989

CONTENTS

Applied Logistic Regression

CHAPTER 1

Introduction to the Logistic Regression Model

1.1 Introduction

Regression methods have become an integral component of any data analysis concerned with describing the relationship between a response variable and one or more explanatory variables. It is often the case that the outcome variable is discrete, taking on two or more possible values. Over the last decade the logistic regression model has become, in many fields, the standard method of analysis in this situation.

Before beginning a study of logistic regression it is important to understand that the goal of an analysis using this method is the same as that of any model-building technique used in statistics: To find the best fitting and most parsimonious, yet biologically reasonable model to describe the relationship between an outcome (dependent or response variable) and a set of independent (predictor or explanatory) variables. These independent variables are often called **covariates**. The most common example of modeling, and one assumed to be familiar to the readers of this text, is the usual linear regression model where the outcome variable is assumed to be continuous.

What distinguishes a logistic regression model from the linear regression model is that the outcome variable in logistic regression is **binary** or **dichotomous**. This difference between logistic and linear regression is reflected both in the choice of a parametric model and in the assumptions. Once this difference is accounted for, the methods employed in an analysis using logistic regression follow the same general principles used in linear regression. Thus, the techniques used in linear regression analysis will motivate our approach to logistic regression. We illustrate both the similarities and differences between logistic regression and linear regression with an example.

Example

Table 1.1 lists age in years (AGE), and presence or absence of evidence of significant coronary heart disease (CHD) for 100 subjects selected to participate in a study. The table also contains an identifier variable (ID) and an age group variable (AGRP). The outcome variable is CHD, which is coded with a value of zero to indicate CHD is absent, or 1 to indicate that it is present in the individual.

It is of interest to explore the relationship between age and the presence or absence of CHD in this study population. Had our outcome variable been continuous rather than binary, we probably would begin by forming a scatterplot of the outcome versus the independent variable. We would use this scatterplot to provide an impression of the nature and strength of any relationship between the outcome and the independent variable. A scatterplot of the data in Table 1.1 is given in Figure 1.1.

In this scatterplot all points fall on one of two parallel lines representing the absence of CHD ($y = 0$) and the presence of CHD ($y = 1$). There is some tendency for the individuals with no evidence of CHD to be younger than those with evidence of CHD. While this plot does depict the dichotomous nature of the outcome variable quite clearly, it does not provide a clear picture of the nature of the relationship between CHD and age.

A problem with Figure 1.1 is that the variability in CHD at all ages is large. This makes it difficult to describe the functional relationship between age and CHD. One common method of removing some variation while still maintaining the structure of the relationship between the outcome and the independent variable is to create intervals for the independent variable and compute the mean of the outcome variable within each group. In Table 1.2 this strategy is carried out by using the age group variable, AGRP, which categorizes the age data of Table 1.1. Table 1.2 contains, for each age group, the frequency of occurrence of each outcome as well as the mean (or proportion with CHD present) for each group.

Table 1.1 Age and Coronary Heart Disease Status (CHD) of 100 Subjects.

ID	AGRP	AGE	CHD	ID	AGRP	AGE	CHD	ID	AGRP	AGE	CHD
1	1	20	0	35	3	38	0	68	6	51	0
2	1	23	0	36	3	39	0	69	6	52	0
3	1	24	0	37	3	39	1	70	6	52	1
4	1	25	0	38	4	40	0	71	6	53	1
5	1	25	1	39	4	40	1	72	6	53	1
6	1	26	0	40	4	41	0	73	6	54	1
7	1	26	0	41	4	41	0	74	7	55	0
8	1	28	0	42	4	42	0	75	7	55	1
9	1	28	0	43	4	42	0	76	7	55	1
10	1	29	0	44	4	42	0	77	7	56	1
11	2	30	0	45	4	42	1	78	7	56	1
12	2	30	0	46	4	43	0	79	7	56	1
13	2	30	0	47	4	43	0	80	7	57	0
14	2	30	0	48	4	43	1	81	7	57	0
15	2	30	0	49	4	44	0	82	7	57	1
16	2	30	1	50	4	44	0	83	7	57	1
17	2	32	0	51	4	44	1	84	7	57	1
18	2	32	0	52	4	44	1	85	7	57	1
19	2	33	0	53	5	45	0	86	7	58	0
20	2	33	0	54	5	45	1	87	7	58	1
21	2	34	0	55	5	46	0	88	7	58	1
22	2	34	0	56	5	46	1	89	7	59	1
23	2	34	1	57	5	47	0	90	7	59	1
24	2	34	0	58	5	47	0	91	8	60	0
25	2	34	0	59	5	47	1	92	8	60	1
26	3	35	0	60	5	48	0	93	8	61	1
27	3	35	0	61	5	48	1	94	8	62	1
28	3	36	0	62	5	48	1	95	8	62	1
29	3	36	1	63	5	49	0	96	8	63	1
30	3	36	0	64	5	49	0	97	8	64	0
31	3	37	0	65	5	49	1	98	8	64	1
32	3	37	1	66	6	50	0	99	8	65	1
33	3	37	0	67	6	50	1	100	8	69	1
34	3	38	0								

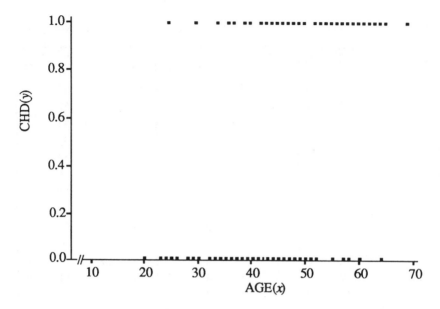

Figure 1.1   Plot of CHD by Age.

By examining this table, a clearer picture of the relationship begins to emerge. It appears that as age increases, the proportion of individuals with evidence of CHD increases. Figure 1.2 presents a plot of the proportion of individuals with CHD versus the midpoint of each age interval. While this

Table 1.2   Frequency Table of Age Group by CHD.

		CHD		
Age Group	n	Absent	Present	Mean (Proportion)
20 – 29	10	9	1	0.10
30 – 34	15	13	2	0.13
35 – 39	12	9	3	0.25
40 – 44	15	10	5	0.33
45 – 49	13	7	6	0.46
50 – 54	8	3	5	0.63
55 – 59	17	4	13	0.76
60 – 69	10	2	8	0.80
Total	100	57	43	0.43

provides considerable insight into the relationship between CHD and age in this study, a functional form for this relationship needs to be described. The plot in this figure is similar to what one might obtain if this same process of grouping and averaging were performed in a linear regression. We will note two important differences.

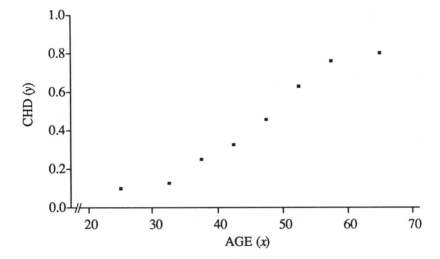

Figure 1.2 Plot of the Mean of CHD in Each Age Group.

The first difference concerns the nature of the relationship between the outcome and independent variable. In any regression problem the key quantity is the mean value of the outcome variable, given the value of the independent variable. This quantity is called the **conditional mean** and will be expressed as "$E(Y|x)$" where Y denotes the outcome variable and x denotes a value of the independent variable. The quantity $E(Y|x)$ is read "the expected value of Y, given the value x." In linear regression we assume that this mean may be expressed as an equation linear in x (or some transformation of x or Y), such as

$$E(Y|x) = \beta_0 + \beta_1 x$$

This expression implies that it is possible for $E(Y|x)$ to take on any value as x ranges between $-\infty$ and $+\infty$.

The column labeled "Mean" in Table 1.2 provides an estimate of $E(Y|x)$. We will assume, for purposes of exposition, that the estimated values plotted in Figure 1.2 are close enough to the true values of $E(Y|x)$ to provide a reasonable assessment of the relationship between CHD and age. With dichotomous data the conditional mean must be greater than or equal to zero and less than or equal to 1 [i.e., $0 \leq E(Y|x) \leq 1$]. This can be seen in Figure 1.2. In addition, the plot shows that this mean approaches zero and 1 "gradually." The change in the $E(Y|x)$ per-unit change in x becomes progressively smaller as the conditional mean gets closer to zero or 1. The curve is said to be **S-shaped**. It resembles a plot of a cumulative distribution of a random variable. It should not seem surprising that some well-known cumulative distributions have been used to provide a model for $E(Y|x)$ in the case when Y is dichotomous. The model we will use is that of the logistic distribution.

Many distribution functions have been proposed for use in the analysis of a dichotomous outcome variable. Cox (1970) discusses some of these. There are two primary reasons for choosing the logistic distribution. These are: (1) from a mathematical point of view, it is an extremely flexible and easily used function, and (2) it lends itself to a biologically meaningful interpretation. A detailed discussion of the interpretation of the model is given in Chapter 3.

In order to simplify notation, we will use the quantity $\pi(x) = E(Y|x)$ to represent the conditional mean of Y given x when the logistic distribution is used. The specific form of the logistic regression model we will use is as follows:

$$\pi(x) = \frac{e^{\beta_0 + \beta_1 x}}{1 + e^{\beta_0 + \beta_1 x}} \tag{1.1}$$

A transformation of $\pi(x)$ that will be central to our study of logistic regression is the **logit transformation**. This transformation is defined, in terms of $\pi(x)$, as follows:

$$g(x) = \ln\left[\frac{\pi(x)}{1 - \pi(x)}\right]$$

$$= \beta_0 + \beta_1 x$$

The importance of this transformation is that $g(x)$ has many of the desirable properties of a linear regression model. The logit, $g(x)$ is linear in its parameters, may be continuous, and may range from $-\infty$ to $+\infty$, depending on the range of x.

The second important difference between the linear and logistic regression models concerns the conditional distribution of the outcome variable. In the linear regression model we assume that an observation of the outcome variable may be expressed as $y = E(Y \mid x) + \varepsilon$. The quantity ε is called the **error** and expresses an observation's deviation from the conditional mean. The most common assumption is that ε follows a normal distribution with mean zero and some variance that is constant across levels of the independent variable. It follows that the conditional distribution of the outcome variable given x will be normal with mean $E(Y \mid x)$, and a variance that is constant. This is not the case with a dichotomous outcome variable. In this situation we may express the value of the outcome variable given x as $y = \pi(x) + \varepsilon$. Here the quantity ε may assume one of two possible values. If $y = 1$ then $\varepsilon = 1 - \pi(x)$ with probability $\pi(x)$, and if $y = 0$ then $\varepsilon = -\pi(x)$ with probability $1 - \pi(x)$. Thus, ε has a distribution with mean zero and variance equal to $\pi(x)[1 - \pi(x)]$. That is, the conditional distribution of the outcome variable follows a binomial distribution with probability given by the conditional mean, $\pi(x)$.

In summary, we have seen that in a regression analysis when the outcome variable is dichotomous:

(1) The conditional mean of the regression equation must be formulated to be bounded between zero and 1. We have stated that the logistic regression model, $\pi(x)$ given in equation (1.1), satisfies this constraint.

(2) The binomial, not the normal, distribution describes the distribution of the errors and will be the statistical distribution upon which the analysis is based.

(3) The principles that guide an analysis using linear regression will also guide us in logistic regression.

1.2 Fitting the Logistic Regression Model

Suppose we have a sample of n independent observations of the pair (x_i, y_i), $i = 1, 2, ..., n$, where y_i denotes the value of a dichotomous outcome variable and x_i is the value of the independent variable for the ith subject. Furthermore, assume that the outcome variable has been coded as zero or 1, representing absence or presence of the characteristic, respectively. This coding for a dichotomous outcome will be used throughout the text. To fit the logistic regression model in equation (1.1) to a set of data requires that we estimate the values of β_0 and β_1, the unknown parameters.

In linear regression the method used most often for estimating unknown parameters is **least squares**. In that method we choose those values of β_0 and β_1 which minimize the sum of squared deviations of the observed values of Y from the predicted values based upon the model. Under the usual assumptions for linear regression the method of least squares yields estimators with a number of desirable statistical properties. Unfortunately, when the method of least squares is applied to a model with a dichotomous outcome the estimators no longer have these same properties.

The general method of estimation that leads to the least squares function under the linear regression model (when the error terms are normally distributed) is called **maximum likelihood**. This method will provide the foundation for our approach to estimation with the logistic regression model. In a very general sense the method of maximum likelihood yields values for the unknown parameters which maximize the probability of obtaining the observed set of data. In order to apply this method we must first construct a function, called the **likelihood function**. This function expresses the probability of the observed data as a function of the unknown parameters. The **maximum likelihood estimators** of these parameters are chosen to be those values which maximize this function. Thus, the resulting estimators are those which agree most closely with the observed data. We now describe how to find these values from the logistic regression model.

If Y is coded as zero or one then the expression for $\pi(x)$ given in equation (1.1) provides (for an arbitrary value of $\beta' = (\beta_0, \beta_1)$, the vector of parameters) the conditional probability that Y is equal to 1 given x. This will be denoted as

$P(Y = 1 | x)$. It follows that the quantity $1 - \pi(x)$ gives the conditional probability that Y is equal to zero given x, $P(Y = 0 | x)$. Thus, for those pairs (x_i, y_i), where $y_i = 1$ the contribution to the likelihood function is $\pi(x_i)$, and for those pairs where $y_i = 0$ the contribution to the likelihood function is $1 - \pi(x_i)$, where the quantity $\pi(x_i)$ denotes the value of $\pi(x)$ computed at x_i. A convenient way to express the contribution to the likelihood function for the pair (x_i, y_i) is through the term

$$\zeta(x_i) = \pi(x_i)^{y_i} [1 - \pi(x_i)]^{1-y_i} \tag{1.2}$$

Since the observations are assumed to be independent, the likelihood function is obtained as the product of the terms given in expression (1.2) as follows:

$$l(\beta) = \prod_{i=1}^{n} \zeta(x_i) \tag{1.3}$$

The principle of maximum likelihood states that we use as our estimate of β the value which maximizes the expression in equation (1.3). However, it is easier mathematically to work with the log of equation (1.3). This expression, the **log likelihood**, is defined as

$$L(\beta) = \ln[l(\beta)] = \sum_{i=1}^{n} \{y_i \ln[\pi(x_i)] + (1 - y_i) \ln[1 - \pi(x_i)]\} \tag{1.4}$$

To find the value of β that maximizes $L(\beta)$ we differentiate $L(\beta)$ with respect to β_0 and β_1 and set the resulting expressions equal to zero. These equations are as follows:

$$\sum_{i=1}^{n} [y_i - \pi(x_i)] = 0 \tag{1.5}$$

and

$$\sum_{i=1}^{n} x_i[y_i - \pi(x_i)] = 0 \qquad (1.6)$$

and are called the **likelihood equations**. In equations (1.5) and (1.6) it is understood that the summation indicated by Σ is over i varying from 1 to n. (The practice of suppressing the index and range of summation, when these are clear, will be followed throughout the text.)

In linear regression the likelihood equations, obtained by differentiating the sum of squared deviations function with respect to ß, are linear in the unknown parameters and thus are easily solved. For logistic regression the expressions in equations (1.5) and (1.6) are nonlinear in β_0 and β_1, and thus require special methods for their solution. These methods are iterative in nature and have been programmed into available logistic regression software. For the moment we need not be concerned about these iterative methods and will view them as a computational detail taken care of for us. The interested reader may see the text by McCullagh and Nelder (1983) for a general discussion of the methods used by most programs. In particular, they show that the solution to equations (1.5) and (1.6) may be obtained using a generalized weighted least squares procedure.

The value of ß given by the solution to equations (1.5) and (1.6) is called the maximum likelihood estimate and will be denoted as $\hat{\text{ß}}$. In general, the use of the symbol $\hat{}$ will denote the maximum likelihood estimate of the respective quantity. For example, $\hat{\pi}(x_i)$ is the maximum likelihood estimate of $\pi(x_i)$. This quantity provides an estimate of the conditional probability that Y is equal to 1, given that x is equal to x_i. As such, it represents the fitted or predicted value for the logistic regression model. An interesting consequence of equation (1.5) is that

$$\sum_{i=1}^{n} y_i = \sum_{i=1}^{n} \hat{\pi}(x_i)$$

That is, the sum of the observed values of y is equal to the sum of the predicted (expected) values. This property will be especially useful in later chapters when we discuss assessing the fit of the model.

As an example, consider the data given in Table 1.1. Use of a logistic regression software package, with continuous variable AGE as the independent variable, produces the output in Table 1.3.

Table 1.3 Results of Fitting the Logistic Regression Model to the Data in Table 1.1.

Variable	Estimated Coefficient	Standard Error	Coeff./SE
AGE	0.111	0.024	4.61
Constant	−5.310	1.134	−4.68

Log-likelihood = −53.677

The maximum likelihood estimates of β_0 and β_1 are thus seen to be $\hat{\beta}_0 = -5.310$ and $\hat{\beta}_1 = 0.111$. The fitted values are given by the equation

$$\hat{\pi}(x) = \frac{e^{-5.31+0.111\times AGE}}{1 + e^{-5.31+0.111\times AGE}} \tag{1.7}$$

and the estimated logit, $\hat{g}(x)$, is given by the equation

$$\hat{g}(x) = -5.31 + 0.111\times AGE \tag{1.8}$$

The log likelihood given in Table 1.3 is the value of equation (1.4) computed using $\hat{\beta}_0$ and $\hat{\beta}_1$.

Two additional columns are present in Table 1.3. One contains estimates of the standard errors of the estimated coefficients and the other the ratios of the estimated coefficients to their estimated standard errors. Use of these quantities will be discussed in the next section.

Following the fitting of the model we should begin to evaluate its adequacy.

1.3 Testing for the Significance of the Coefficients

In practice, the modeling of a set of data, as we show in Chapters 4, 7, and 8, is a much more complex process than one of fitting and testing. The methods we present in this section, while simplistic from the point of view of the total picture, do provide essential building blocks for the more complex process.

After estimating the coefficients, our first look at the fitted model commonly concerns an assessment of the significance of the variables in the model. This usually involves formulation and testing of a statistical hypothesis to determine whether the independent variables in the model are "significantly" related to the outcome variable. The method for performing this test is quite general and differs from one type of model to the next only in the specific details. We begin by discussing the general approach for a single independent variable. The multivariate case will be discussed in Chapter 2.

One approach to testing for the significance of the coefficient of a variable in any model relates to the following question. *Does the model that includes the variable in question tell us more about the outcome (or response) variable than does a model that does not include that variable?* This question is answered by comparing the observed values of the response variable to those predicted by each of two models; the first with and the second without the variable in question. The mathematical function used to compare the observed and predicted values depends on the particular problem. If the predicted values with the variable in the model are better, or more accurate in some sense, than when the variable is not in the model, then we feel that the variable in question is "significant." It is important to note that we are not considering the question of whether the predicted values are an accurate representation of the observed values in an absolute sense (this would be called **goodness-of-fit**). Instead, our question is posed in a relative sense. The assessment of goodness-of-fit is a more complex question which will be discussed in detail in Chapter 5.

The general method for assessing significance of variables is easily illustrated in the linear regression model, and its use there will motivate the approach used for logistic regression. A comparison of the two approaches will highlight the differences between modeling a continuous versus a dichotomous response variable.

In linear regression the assessment of the significance of the slope coefficient is approached by first forming what is referred to as an **analysis of variance table.** This table partitions the total sum of squared deviations of observations about their mean into two parts: (1) the sum of squared deviations of observations about the regression line SSE, (or *residual sum-of-squares*), and (2) the sum of squares of predicted values, based on the regression model, about the mean of the dependent variable SSR, (or *due regression sum-of-squares*). This is just a convenient way of displaying the comparison of observed to predicted values under two models. In linear regression the comparison of observed to predicted values is based on the square of the distance between the two. If y_i denotes the observed value and $\hat{y}_i$ denotes the predicted value for the ith individual under the model, then the statistic used to evaluate this comparison is

$$SSE = \sum_{i=1}^{n}(y_i - \hat{y}_i)^2$$

Under the model not containing the independent variable in question the only parameter is β_0, and $\hat{\beta}_0 = \bar{y}$, the mean of the response variable. In this case $\hat{y}_i = \bar{y}$ and SSE is equal to the total variance. When we include the independent variable in the model any decrease in SSE will be due to the fact that the slope coefficient for the independent variable is not zero. The change in the value of SSE is the due to the regression source of variability, denoted SSR. That is,

$$SSR = \left[\sum_{i=1}^{n}(y_i - \bar{y})^2\right] - \left[\sum_{i=1}^{n}(y_i - \hat{y}_i)^2\right]$$

In linear regression, interest focuses on the size of SSR. A large value suggests that the independent variable is important, whereas a small value suggests that the independent variable is not helpful in predicting the response.

The guiding principle with logistic regression is the same: *Compare observed values of the response variable to predicted values obtained from models with and without the variable in question.* In logistic regression comparison of observed to predicted values is based on the log likelihood function defined in equation (1.4). To better understand this comparison, it is helpful conceptually if we think of an observed value of the response variable as also being a predicted value resulting from a **saturated model**. A saturated model is one that contains as many parameters as there are data points. (A simple example of a saturated model is fitting a linear regression model when there are only two data points, $n = 2$.)

The comparison of observed to predicted values using the likelihood function is based on the following expression:

$$D = -2 \ln\left[\frac{\text{(likelihood of the current model)}}{\text{(likelihood of the saturated model)}} \right] \qquad (1.9)$$

The quantity inside the large brackets in the expression above is called the **likelihood ratio**. The reason for using minus twice its log is mathematical and is necessary to obtain a quantity whose distribution is known and thus can be used for hypothesis testing purposes. Such a test is called the **likelihood ratio test**. Using equation (1.4) equation (1.9) becomes

$$D = -2 \sum_{i=1}^{n}\left[y_i \ln\left(\frac{\hat{\pi}_i}{y_i} \right) + (1 - y_i) \ln\left(\frac{1 - \hat{\pi}_i}{1 - y_i} \right) \right] \qquad (1.10)$$

where $\hat{\pi}_i = \hat{\pi}(x_i)$.

The statistic, D, in equation (1.10) is called the **deviance** by some authors [see, for example, McCullagh and Nelder (1983)], and plays a central role in some approaches to assessment of goodness-of-fit. The deviance for logistic regression plays the same role as the residual sum of squares plays in linear regression. In fact, the deviance as shown in equation (1.9), when computed for linear regression, is identically equal to SSE.

For purposes of assessing the significance of an independent variable we compare the value of D with and without the independent variable in the equation. The change in D due to including the independent variable in the model is obtained as follows:

$G = D(\text{for the model without the variable}) - D(\text{for the model with the variable})$

This statistic plays the same role in logistic regression as does the numerator of the partial F test in linear regression. Because the likelihood of the saturated model is common to both values of D being differenced to compute G, it can be expressed as

$$G = -2 \ln \left[\frac{\text{(likelihood without the variable)}}{\text{(likelihood with the variable)}} \right] \quad (1.11)$$

For the specific case of a single independent variable it is easy to show that when that variable is not in the model, the maximum likelihood estimate of β_0 is $\ln(n_1/n_0)$ where $n_1 = \Sigma y_i$ and $n_0 = \Sigma(1 - y_i)$ and that the predicted value is constant, n_1/n In this case the value of G is as follows:

$$G = -2 \ln \left[\frac{\left(\frac{n_1}{n}\right)^{n_1} \left(\frac{n_0}{n}\right)^{n_0}}{\displaystyle\prod_{i=1}^{n} \hat{\pi}_i^{y_i} (1 - \hat{\pi}_i)^{(1-y_i)}} \right] \quad (1.12)$$

or,

$$G = 2 \left\{ \sum_{i=1}^{n} [y_i \ln(\hat{\pi}_i) + (1-y_i) \ln(1-\hat{\pi}_i)] - [n_1 \ln(n_1) + n_0 \ln(n_0) - n \ln(n)] \right\}$$

$$(1.13)$$

Under the hypothesis that β_1 is equal to zero, the statistic G will follow a chi-square distribution with 1 degree of freedom. Additional mathematical assumptions are also needed; but for the above case they are rather nonrestrictive and involve having a sufficiently large sample size, n.

As an example we consider the model fit to the data in Table 1.1, whose estimated coefficients and log likelihood are given in Table 1.3. For these data $n_1 = 43$ and $n_0 = 57$; thus, evaluating G as shown in equation (1.13) yields

$$G = 2\{-53.677 - [43 \ln(43) + 57 \ln(57) - 100 \ln(100)]\}$$

$$= 2[-53.677 - (68.322)]$$

$$= 29.31$$

The first term in this expression is the log likelihood with the variable from Table 1.3, and the remainder of the expression simply substitutes n_1 and n_0 into the second part of equation (1.13). We will use the symbol $\chi^2(v)$ to denote a chi-square random variable with v degrees of freedom. Using this notation, the p-value associated with this test is $P[\chi^2(1) > 29.31] < 0.001$; thus we have convincing evidence that AGE is a significant variable in predicting CHD. This is merely a statement of the statistical evidence for this variable. Other important factors to consider before concluding that the variable is biologically important would include the appropriateness of the fitted model, as well as inclusion of other potentially important variables.

The calculation of the log likelihood and the likelihood ratio test are standard features of any good logistic regression package. This makes it possible to check for the significance of the addition of new terms to the model as a matter of routine. In the simple case of a single independent variable, we can first fit a model containing only the constant term. We can then fit a model containing the independent variable along with the constant. This gives rise to a new log likelihood. The likelihood ratio test is obtained by multiplying the difference in these two values by -2.

In the current example, the log likelihood for the model containing only a constant term is -68.332. Fitting a model containing the independent variable, AGE, along with the constant term results in a log likelihood shown in Table 1.3 of -53.677. Multiplying the difference in these log likelihoods by -2 gives

$$-2 \times [-68.332 - (-53.677)] = -2 \times (-14.655) = 29.31$$

This result, along with the associated p-value for the chi-square distribution, may be obtained from most software packages.

Two other similar, statistically equivalent tests have been suggested. These are the Wald test and Score test. The assumptions needed for these tests are the same as those of the likelihood ratio test in equation (1.12). A more complete discussion of these tests and their assumptions may be found in Rao (1973).

The Wald test is obtained by comparing the maximum likelihood estimate of the slope parameter, $\hat{\beta}_1$, to an estimate of its standard error. The resulting ratio, under the hypothesis that $\beta_1 = 0$, will follow a standard normal distribution. While we have not yet formally discussed how the estimates of the standard errors of the estimated parameters are obtained, they are routinely printed out by computer software. For example, the Wald test for the logistic regression model in Table 1.3 is

$$W = \frac{\hat{\beta}_1}{\hat{SE}(\hat{\beta}_1)} = \frac{0.111}{0.024} = 4.610$$

and the two tailed p-value is $P(|Z| > 4.610)$, where Z denotes a random variable following the standard normal distribution. Hauck and Donner (1977) examined the performance of the Wald test and found that it behaved in an aberrant manner, often failing to reject when the coefficient was significant. They recommended that the likelihood ratio test be used.

Jennings (1986a) has also looked at the adequacy of inferences in logistic regression based on Wald statistics. His conclusions are similar to those of Hauck and Donner. Both the likelihood ratio test, G, and the Wald test, W, require the computation of the maximum likelihood estimate for β_1. For a single variable this is not a difficult or costly computational task. However, for large data sets with many variables, the iterative computation needed to obtain the maximum likelihood estimates can be considerable.

A test for the significance of a variable which does not require these computations is the Score test. Proponents of the Score test cite this reduced computational effort as its major advantage. Use of the test is limited by the fact that it cannot be obtained easily from some software packages. The Score test is based on the distribution theory of the derivatives of the log likelihood. In general, this is a multivariate test requiring matrix calculations which will be discussed in Chapter 2.

In the univariate case this test is based on the conditional distribution of the derivative in equation (1.6), given the derivative in equation (1.5). In this case, we can write down an expression for the Score test. The test uses the value of equation (1.6), computed using $\beta_0 = \ln(n_1/n_0)$ and $\beta_1 = 0$. As noted earlier, under these parameter values, $\hat{\pi} = n_1/n = \bar{y}$. Thus, the left-hand side of equation (1.6) becomes $\sum x_i(y_i - \bar{y})$. It may be shown that the estimated variance is $\bar{y}(1 - \bar{y})\sum(x_i - \bar{x})^2$. The test statistic for the Score test (ST) is

$$ST = \frac{\displaystyle\sum_{i=1}^{n} x_i(y_i - \bar{y})}{\sqrt{\bar{y}(1 - \bar{y})\displaystyle\sum_{i=1}^{n}(x_i - \bar{x})^2}}$$

As an example of the Score test consider the model fit to the data in Table 1.1. The value of the test statistic for this example is

$$ST = 296.66/(3333.742)^{1/2}$$

$$= 5.14$$

and the two tailed p-value is $P(|Z| > 5.14) < 0.001$. We note that, for this example, the values of the three test statistics are nearly the same (note: $\sqrt{G} = 5.41$).

In summary, the method for testing the significance of the coefficient of a variable in logistic regression follows a similar approach to that used in linear regression, but uses the likelihood function for a dichotomous outcome variable.

1.4 Other Methods of Estimation

The method of maximum likelihood described in Section 1.2 is the estimation method used in the logistic regression routines of the major software packages. However, two other methods have been and may still be used for estimating the coefficients. These methods are: (1) noniterative weighted least squares, and (2) discriminant function analysis.

A linear models approach to the analysis of categorical data was proposed by Grizzle, Starmer, and Koch (1969), which uses estimators based on noniterative weighted least squares. They demonstrate that the logistic regression model is an example of a very general class of models that can be handled with their methods. We should add that the maximum likelihood estimators are usually calculated using an iterative reweighted least squares algorithm, and thus are also "least squares" estimators. The approach suggested by Grizzle et al. uses only one iteration in the process.

A major limitation of this method is that we must have an estimate of $\pi(x)$ which is not zero or 1 for most values of x. An example where we could use both maximum likelihood and noniterative weighted least squares is the data in Table 1.2. In cases such as this, the two methods are **asymptotically equivalent**. The term asymptotically equivalent means that as n gets large, the distributional properties of the estimators become identical.

The discriminant function approach to estimation of the coefficients was popularized by Cornfield (1962) in his early work on logistic regression. These estimators take their name from the fact that the posterior probability in the usual discriminant function model is the logistic regression function given in equation (1.1). More precisely, if the independent variable, X, follows a normal distribution within each of two groups (subpopulations) defined by the two values of y having different means and the same variance, then the conditional distribution of Y given $X = x$ is the logistic regression model. That is, if $X \mid Y = j \sim N(\mu_j, \sigma^2)$, $j = 0, 1$ then $P(Y = 1 \mid x) = \pi(x)$. The symbol "$\sim$" is read "is distributed" and the "$N(\mu, \sigma^2)$" denotes the normal distribution with mean equal to μ and variance equal to σ^2. Under these assumptions it is easy to show [Lachenbruch (1975)] that the logistic coefficients are

$$\beta_0 = \ln\left(\frac{\theta_1}{\theta_0}\right) - 0.5(\mu_1 - \mu_0)^2/\sigma^2 \tag{1.14}$$

and

$$\beta_1 = (\mu_1 - \mu_0)/\sigma^2 \tag{1.15}$$

where $\theta_j = P(Y = j)$, $j = 0, 1$. The discriminant function estimators of β_0 and β_1 are found by substituting estimators for μ_j, θ_j, $j = 0, 1$ and σ^2 into the above equations. The estimators usually used are $\hat{\mu}_j = \bar{x}_j$, the mean of x in the subgroup defined by $y = j$, $j = 0, 1$, $\theta_1 = n_1/n$ the mean of y with

$$\hat{\theta}_0 = 1 - \hat{\theta}_1$$

and

$$\hat{\sigma}^2 = \left[(n_0 - 1)s_0^2 + (n_1 - 1)s_1^2\right]\big/(n_0 + n_1 - 2)$$

where s_j^2 is the unbiased estimator of σ^2 computed within the subgroup of the data defined by $y = j, j = 0, 1$. The above expressions are for a single variable x; the multivariate case will be presented in Chapter 2.

It is natural to ask why, if the discriminant function estimators are so easy to compute, are they not used in place of the maximum likelihood estimators? Halpern, Blackwelder, and Verter (1971) and Hosmer, Hosmer, and Fisher (1983) have compared the two methods when the model contains a mixture of continuous and discrete variables, with the general conclusion that the discriminant function estimators are sensitive to the assumption of normality. In particular, the estimators of the coefficients for nonnormally distributed variables are biased away from zero when the coefficient is, in fact, different from zero. The practical implication of this is that for dichotomous independent variables (and we will see that this is a commonly occurring independent variable in many situations), the discriminant function estimators will overestimate the magnitude of the association.

At this point it may be helpful to more carefully delineate the various uses of the term "maximum likelihood," as it applies to the estimation of the logistic regression coefficients. Under the assumptions of the discriminant function model stated above, the estimators obtained from equations (1.14) and (1.15) are maximum likelihood estimators. Those obtained from equations (1.5) and (1.6) are based on the conditional distribution of Y given X and, as such, are actually "conditional maximum likelihood estimators." Because discriminant function estimators are rarely used anymore, the word conditional has been dropped when describing the estimators given in equations (1.5) and (1.6). We will use the word **conditional** to describe estimators in logistic regression with matched data as discussed in Chapter 7.

In summary there are alternative methods of estimation for some configurations of the data which are computationally quicker; but we will use the method of maximum likelihood described in Section 1.2 throughout the rest of this text.

1.5 The ICU Study

A data set which will be used in exercises throughout the text consists of a sample of 200 subjects who were part of a much larger study on survival of patients following admission to an adult intensive care unit (ICU). The major goal of this study was to develop a logistic regression model to predict the probability of survival to hospital discharge of these patients. A number of publications have appeared which have focused on various facets of the problem. The reader wishing to learn more about the clinical aspects of this study should start with Lemeshow, Teres, Avrunin, and Pastides (1988). A code sheet for the variables to be considered in this text is given below in Table 1.4. A listing of the data is provided in Appendix 2.

Table 1.4   Code Sheet for the ICU Data.

Variable	Name	Codes/Values	Column Heading Appendix 2
1	Identification Code	ID Number	ID
2	Vital Status	0 = Lived 1 = Died	STA
3	Age	Years	AGE
4	Sex	0 = Male 1 = Female	SEX
5	Race	1 = White 2 = Black 3 = Other	RACE
6	Service at ICU Admission	0 = Medical 1 = Surgical	SER
7	Cancer Part of Present Problem	0 = No 1 = Yes	CAN
8	History of Chronic Renal Failure	0 = No 1 = Yes	CRN

Table 1.4 Continued.

Variable	Name	Codes/Values	Column Heading Appendix 2
9	Infection Probable at ICU Admission	0 = No 1 = Yes	INF
10	CPR Prior to ICU Admission	0 = No 1 = Yes	CPR
11	Systolic Blood Pressure at ICU Admission	mm Hg	SYS
12	Heart Rate at ICU Admission	Beats/min	HRA
13	Previous Admission to an ICU within 6 Months	0 = No 1 = Yes	PRE
14	Type of Admission	0 = Elective 1 = Emergency	TYP
15	Long Bone, Multiple, Neck, Single Area, or Hip Fracture	0 = No 1 = Yes	FRA
16	PO2 from Initial Blood Gases	0 = >60 1 = 60	PO2
17	PH from Initial Blood Gases	0 = 7.25 1 = <7.25	PH
18	PCO2 from Initial Blood Gases	0 = 45 1 = >45	PCO
19	Bicarbonate from InitialBlood Gases	0 = 18 1 = <18	BIC
20	Creatinine from Initial Blood Gases	0 = 2.0 1 = >2.0	CRE
21	Level of Consciousness at ICU Admission	0 = No Coma or Stupor 1 = Deep Stupor 2 = Coma	LOC

Exercises

1. In the ICU data described in Section 1.5 the primary outcome variable is vital status at hospital discharge, STA. Clinicians associated with the study felt that a key determinant of survival was the patient's age at admission, AGE.

    1.1 Write down the equation for the logistic regression model of STA on AGE. Write down the equation for the logit transformation of this logistic regression model. What characteristic of the outcome variable, STA, leads us to consider the logistic regression model as opposed to the usual linear regression model to describe the relationship between STA and AGE?

    1.2 Form a scatterplot of STA versus AGE.

    1.3 Using intervals based on the empirical octiles (eighths) of AGE, compute the STA mean over subjects within each AGE interval. Plot these values of mean STA versus the midpoint of the AGE interval using the same set of axes as was used in problem 1.2.

    1.4 Write down an expressions for the likelihood and log-likelihood for the logistic regression model in problem 1.1 using the ungrouped, $n = 200$, data. Obtain expressions for the two likelihood equations.

    1.5 Using a logistic regression package of your choice obtain the maximum likelihood estimates of the parameters of the logistic regression model in problem 1.1. These estimates should be based on the ungrouped, $n = 200$, data. Using these estimates, write down the equation for the fitted values, that is, the estimated logistic probabilities. Plot the equation for the fitted values on the axes used in the scatterplots in problems 1.2 and 1.3.

    1.6 Summarize (describe in words) the results presented in the plot obtained from problems 1.2, 1.3, and 1.5.

    1.7 Using the results of the output from the logistic regression package used for problem 1.4, assess the significance of the slope coefficient for AGE using the likelihood ratio test, the Wald test, and, if possible, the Score test. What assumptions are needed for the p-values computed for each of these tests to be valid? Are the results of these tests consistent with one another? What is the value of the deviance for the fitted model?

1.8 Compute the values of discriminant function estimates of the parameters in the logistic regression model of STA on AGE and compare them to the estimates obtained in problem 1.4. Briefly summarize the assumptions necessary for the discriminant function estimators to be valid and compare them to the assumptions necessary for the conditional maximum likelihood estimators used in problem 1.3.

2. Repeat problems 1.1, 1.2, and 1.4-1.8 using the variable "type of admission," TYP, as the covariate. The variable TYP is dichotomous so the scatterplot is actually a 2×2 contingency table.

CHAPTER 2

The Multiple Logistic Regression Model

2.1 Introduction

In the previous chapter we introduced the logistic regression model in the univariate context. As in the case of linear regression, the strength of a modeling technique lies in its ability to model many variables, some of which may be on different measurement scales. In this chapter we will generalize the logistic model to the case of more than one independent variable. This will be referred to as the "multivariate case." Central to the consideration of the multiple logistic model will be estimation of the coefficients in the model and testing for their significance. This will follow along the same lines as the univariate model. An additional modeling consideration which will be introduced in this chapter is the use of design variables for modeling discrete, nominal scaled independent variables. In all cases it will be assumed that there is a predetermined collection of variables which is being examined. The question of variable selection will be dealt with in Chapter 4. An example is used to illustrate the multivariate model, as well as the estimation and testing of the coefficients.

2.2 The Multiple Logistic Regression Model

Consider a collection of p independent variables which will be denoted by the vector $\mathbf{x}' = (x_1, x_2, ..., x_p)$. For the moment we will assume that each of these variables is at least interval scaled. Let the conditional probability that the outcome is present be denoted by $P(Y = 1 \mid \mathbf{x}) = \pi(\mathbf{x})$. Then the logit of the multiple logistic regression model is given by the equation

$$g(\mathbf{x}) = \beta_0 + \beta_1 x_1 + \beta_2 x_2 + \cdots + \beta_p x_p \tag{2.1}$$

in which case

$$\pi(\mathbf{x}) = \frac{e^{g(\mathbf{x})}}{1 + e^{g(\mathbf{x})}} \tag{2.2}$$

If some of the independent variables are discrete, nominal scaled variables such as race, sex, treatment group, and so forth, then it is inappropriate to include them in the model as if they were interval scaled. This is because the numbers used to represent the various levels are merely identifiers, and have no numeric significance. In this situation the method of choice is to use a collection of **design variables** (or **dummy variables**). Suppose, for example, that one of the independent variables is race, which has been coded as "white," "black" or "other." In this case two design variables are necessary. One possible coding strategy is that when the respondent is "white," the two design variables, D_1 and D_2, would both be set equal to zero; when the respondent is "black," D_1 would be set equal to 1 while D_2 would still equal 0; when the race of the respondent is "other," we would use $D_1 = 0$ and $D_2 = 1$. Table 2.1 illustrates this coding of the design variables.

Table 2.1   An Example of the Coding of the Design Variables for Race, Coded at Three Levels.

	Design Variable	
RACE	D_1	D_2
White	0	0
Black	1	0
Other	0	1

Most logistic regression software will generate the design variables, and some programs have a choice of several different methods. The different strategies for creation and interpretation of design variables will be discussed in detail in Chapter 3.

In general, if a nominal scaled variable has k possible values, then $k - 1$ design variables will be needed. This is true since, unless stated otherwise, all of our models have a constant term. The notation to indicate design variables to be used in this text follows. Suppose that the jth independent variable, x_j has k_j levels. The $k_j - 1$ design variables will be denoted as D_{ju} and the coefficients for these design variables will be denoted as β_{ju}, $u = 1, 2, ..., k_j - 1$. Thus, the logit for a model with p variables and the jth variable being discrete would be

$$g(\mathbf{x}) = \beta_0 + \beta_1 x_1 + \cdots + \sum_{u=1}^{k_j-1} \beta_{ju} D_{ju} + \beta_p x_p.$$

When discussing the multiple logistic regression model we will, in general, suppress the summation and double subscripting needed to indicate when design variables are being used. The exception to this will be the discussion of modeling strategies when we will need to use the specific value of the coefficients for any design variables in the model.

2.3 Fitting the Multiple Logistic Regression Model

Assume that we have a sample of n independent observations of the pair $(\mathbf{x}_i, y_i)$, $i = 1, 2, ..., n$. As in the univariate case, fitting the model requires that we obtain estimates of the vector $\boldsymbol{\beta}' = (\beta_0, \beta_1, ..., \beta_p)$. The method of estimation used in the multivariate case will be the same as in the univariate situation – maximum likelihood. The likelihood function is nearly identical to that given in equation (1.3), with the only change being that $\pi(\mathbf{x})$ is now defined as in equation (2.2). There will be $p + 1$ likelihood equations which are obtained by differentiating the log likelihood function with respect to the $p + 1$ coefficients. The likelihood equations that result may be expressed as follows:

$$\sum_{i=1}^{n} [y_i - \pi(\mathbf{x}_i)] = 0$$

and

$$\sum_{i=1}^{n} x_{ij} [y_i - \pi(\mathbf{x}_i)] = 0$$

for $j = 1, 2, ..., p$.

As in the univariate model, the solution of the likelihood equations requires special purpose software which may be found in many packaged programs. Let

$\hat{\beta}$ denote the solution to these equations. Thus, the fitted values for the multiple logistic regression model are $\hat{\pi}(x_i)$, the value of the expression in equation (2.2) computed using $\hat{\beta}$, and x_i.

In the previous chapter only a brief mention was made of the method for estimating the standard errors of the estimated coefficients. Now that the logistic regression model has been generalized both in concept and notation to the multivariate case, we will consider estimation of standard errors in more detail.

The method of estimating the variances and covariances of the estimated coefficients follows from well-developed theory of maximum likelihood estimation [see, for example, Rao (1973)]. This theory states that the estimators are obtained from the matrix of second partial derivatives of the log likelihood function. These partial derivatives have the following general form

$$\frac{\partial^2 L(\beta)}{\partial \beta_j^2} = -\sum_{i=1}^{n} x_{ij}^2 \pi_i (1 - \pi_i) \qquad (2.3)$$

and

$$\frac{\partial^2 L(\beta)}{\partial \beta_j \partial \beta_u} = -\sum_{i=1}^{n} x_{ij} x_{iu} \pi_i (1 - \pi_i) \qquad (2.4)$$

for $j, u = 0, 1, 2, ..., p$ where π_i denotes $\pi(x_i)$. Let the $(p + 1)$ by $(p + 1)$ matrix containing the negative of the terms given in equations (2.3) and (2.4) be denoted as $I(\beta)$. This matrix is called the **information matrix**. The variances and covariances of the estimated coefficients are obtained from the inverse of this matrix which we will denote as $\Sigma(\beta) = I^{-1}(\beta)$. Except in very special cases it is not possible to write down an explicit expression for the elements in this matrix. Hence, we will use the notation $\sigma^2(\beta_j)$ to denote the jth diagonal element of this matrix, which is the variance of $\hat{\beta}_j$, and $\sigma(\beta_j, \beta_u)$ to denote an arbitrary off-diagonal element, which is the covariance of $\hat{\beta}_j$ and $\hat{\beta}_u$. The estimators of the variances and covariances, which will be denoted by $\hat{\Sigma}(\hat{\beta})$, are obtained by evaluating $\Sigma(\beta)$ at $\hat{\beta}$. We will use $\hat{\sigma}^2(\hat{\beta}_j)$ and $\hat{\sigma}(\hat{\beta}_j, \hat{\beta}_u)$, $j, u = 0, 1, 2, ..., p$, to denote the values in this matrix. For the most part we will have occasion to use only the estimated standard errors of the estimated coefficients, which we will denote as

$$\hat{\text{SE}}(\hat{\beta}_j) = [\hat{\sigma}^2(\hat{\beta}_j)]^{1/2} \tag{2.5}$$

for $j = 0, 1, 2, ..., p$. We will use this notation in developing methods for coefficient testing and confidence interval estimation.

A formulation of the information matrix which will be useful when discussing model fitting and assessment of fit is $\hat{\mathbf{I}}(\hat{\beta}) = \mathbf{X}'\mathbf{V}\mathbf{X}$ where $\mathbf{X}$ is an n by $p + 1$ matrix containing the data for each subject, and $\mathbf{V}$ is an n by n diagonal matrix with general element $\hat{\pi}_i(1 - \hat{\pi}_i)$. That is, the matrix $\mathbf{X}$ is

$$\mathbf{X} = \begin{bmatrix} 1 & x_{11} & \cdots & x_{1p} \\ 1 & x_{21} & \cdots & x_{2p} \\ & & \vdots & \\ 1 & x_{n1} & \cdots & x_{np} \end{bmatrix}$$

and the matrix $\mathbf{V}$ is

$$\mathbf{V} = \begin{bmatrix} \hat{\pi}_1(1 - \hat{\pi}_1) & 0 & \cdots & 0 \\ 0 & \hat{\pi}_2(1 - \hat{\pi}_2) & \cdots & 0 \\ & & \vdots & \\ 0 & 0 & \cdots & \hat{\pi}_n(1 - \hat{\pi}_n) \end{bmatrix}$$

Before proceeding further we present an example that will illustrate the formulation of a multiple logistic regression model and the estimation of its coefficients.

Example

To provide an example of fitting a multiple logistic regression model, we will use a subset of the variables from the data for the low birth weight study described in Appendix 1. The code sheet for the full data set is given in Table 4.1. The goal of this study was to identify risk factors associated with giving birth to a low birth weight baby (weighing less than 2500 grams). In this study data were collected on 189 women, $n_1 = 59$ of which had low birth weight babies and $n_0 = 130$ of which had normal birth weight babies. Four variables which were thought to be of importance were age, weight of the subject at her last menstrual period, race, and number of physician visits during the first trimester of the pregnancy. In this example, the variable race has been recoded using the two design variables shown in Table 2.1. The results of fitting the logistic regression model to these data are given in Table 2.2.

Table 2.2 Estimated Coefficients for a Multiple Logistic Regression Model Using the Variables AGE, Weight at Last Menstrual Period (LWT), RACE, and Number of First Trimester Physician Visits (FTV) from the Low Birth Weight Data Set.

Variable	Estimated Coefficient	Standard Error	Coeff./SE
AGE	−0.024	0.034	−0.71
LWT	−0.014	0.652E−02	−2.14
RACE (1)	1.004	0.497	2.02
RACE (2)	0.433	0.362	1.20
FTV	−0.049	0.167	−0.30
Constant	1.295	1.069	1.21

Log-likelihood = −111.286

In Table 2.2 the estimated coefficients for the two design variables for race are indicated in the lines denoted by "(1)" and "(2)." The estimated logit is given by the following expression:

$$\hat{g}(\mathbf{x}) = 1.295 - 0.024 \times \text{AGE} - 0.014 \times \text{LWT} + 1.004 \times D_{31}$$
$$+ 0.433 \times D_{32} - 0.049 \times \text{FTV}$$

where D_{3i}, $i = 1, 2$, denotes the two design variables for RACE. The fitted values are obtained using the estimated logit, $\hat{g}(\mathbf{x})$.

2.4 Testing for the Significance of the Model

Once we have fit a particular multiple (multivariate) logistic regression model, we begin the process of assessment of the model. As in the univariate case presented in Chapter 1, the first step in this process is usually assessing the significance of the variables in the model. The likelihood ratio test for overall significance of the p coefficients for the independent variables in the model is performed in exactly the same manner as in the univariate case. The test is based on the statistic G given in equations (1.10) and (1.11). The only difference is that the fitted values, $\hat{\pi}$, under the model are based on the vector containing $p + 1$ parameters, $\hat{\beta}$. Under the null hypothesis that the p "slope" coefficients for the covariates in the model are equal to zero, the distribution of G will be chi-square with p degrees of freedom.

As an example, consider the fitted model whose estimated coefficients are given in Table 2.2. For that model the value of the log likelihood is $L = -111.286$. A second model, fit with the constant term only, yields $L = -117.336$. Hence $G = -2[(-117.336) - (-111.286)] = -2(-6.05) = 12.1$. Alternatively, using equation (1.13) we have, with $n_1 = 59$ and $n_0 = 130$,

$$G = 2\{(-111.286) - [59 \ln(59) + 130 \ln(130) - 189 \ln(189)]\}$$

$$= 12.099$$

and the p-value for the test is $P[\chi^2(5) > 12.099] = 0.033$ which is significant at the $\alpha = 0.05$ level. Rejection of the null hypothesis in this case has an interpretation analogous to that in multiple linear regression; we may conclude that at least one, and perhaps all p coefficients are different from zero.

Before concluding that any or all of the coefficients are nonzero, we may wish to look at the univariate Wald test statistics, $W_j = \hat{\beta}_j / \hat{SE}(\hat{\beta}_j)$. These are given in the last column in Table 2.2. Under the hypothesis that an individual coefficient is zero, these statistics will follow the standard normal distribution. Thus, the value of these statistics may give us an indication of which of the variables in the model may or may not be significant. If we use a critical value of 2, which would lead to an approximate level of significance of 0.05, then we would conclude that the variables LWT and possibly RACE are significant, while AGE and FTV are not significant.

Considering that the overall goal is to obtain the best fitting model while minimizing the number of parameters, the next logical step is to fit a reduced model containing only those variables thought to be significant, and compare it to the full model containing all the variables. The results of fitting the reduced model are given in Table 2.3.

Table 2.3 Estimated Coefficients for a Multiple Logistic Regression Model Using the Variables LWT and RACE from the Low Birth Weight Data Set.

Variable	Estimated Coefficient	Standard Error	Coeff./SE
LWT	−0.015	0.642E-02	−2.37
RACE (1)	1.081	0.487	2.22
RACE (2)	0.481	0.356	1.35
Constant	0.806	0.843	0.96

Log–likelihood = −111.630

The difference between the two models is the exclusion of the variables AGE and FTV from the full model. The likelihood ratio test comparing these two models is obtained using the definition of G given in equation (1.11). It will have a distribution that is chi-square with 2 degrees of freedom under the hypothesis that the coefficients for the variables excluded are equal to zero. The value of the test statistic comparing the models in Tables 2.2 and 2.3 is

$$G = -2[(-111.630) - (-111.286)] = 0.688$$

which, with 2 degrees of freedom, has a p-value of $P[\chi^2(2) > 0.688] = 0.71$. Since the p-value is large, exceeding 0.05, we conclude that the reduced model is as good as the full model. Thus there is no advantage to including AGE and FTV in the model. However, we must not base our models entirely on tests of statistical significance. As we will see in Chapter 5, there are numerous other considerations that will influence our decision to include or exclude variables from a model.

Whenever a categorical scaled independent variable is included (or excluded) from a model, all of its design variables should be included (or excluded); to do otherwise implies that we have recoded the variable. For example, if we only include design variable D_1 as defined in Table 2.1, then race is entered into the model as a dichotomous variable coded as black or not black. If k is the number of levels of a categorical variable, then the contribution to the degrees of freedom for the likelihood ratio test for the exclusion of this variable will be $k - 1$. For example, if we exclude race from the model, and race is coded at three levels using the design variables shown in Table 2.1, then there would be 2 degrees of freedom for the test, one for each design variable.

Because of the multiple degrees of freedom we must be careful in our use of the Wald (W) statistics to assess the significance of the coefficients. For example, if the W statistics for both coefficients exceed 2, then we could conclude that the design variables are significant. Alternatively, if one coefficient has a W statistic of 3.0 and the other a value of 0.1, then we cannot be sure about the contribution of the variable to the model. The estimated coefficients for the variable RACE in Table 2.3 provide a good example. The Wald statistic for the coefficient for the first design variable is 2.22, and 1.35 for the second. The likelihood ratio test comparing the model containing LWT and RACE to the one containing only LWT yields $G = -2[-114.345 - (-111.630)]$ $= 5.43$ which, with 2 degrees of freedom, yields a p-value of 0.066. Strict adherence to the $\alpha = 0.05$ level of significance would justify excluding RACE from the model. However, RACE is known to be a "biologically important" variable. In this case the decision to include or exclude RACE should be made in conjunction with subject matter experts.

In the previous chapter we described, for the univariate model, two other tests equivalent to the likelihood ratio test for assessing the significance of the model; these are the Wald and Score tests. We will briefly discuss the multivariate versions of these tests, as their use appears occasionally in the literature. Application of the Wald or Score tests requires either special software or a computationally sophisticated user. We will only use likelihood ratio-based tests in this text. As noted earlier, we favor the likelihood ratio test as the quantities needed to carry it out may be obtained from all computer packages.

The multivariate analog of the Wald test is obtained from the following vector–matrix calculation

$$W = \hat{\beta}'[\hat{\Sigma}(\hat{\beta})]^{-1}\hat{\beta}$$

$$= \hat{\beta}'(\mathbf{X}'\mathbf{V}\mathbf{X})\hat{\beta}$$

which will be distributed as chi-square with $p + 1$ degrees of freedom under the hypothesis that each of the $p + 1$ coefficients is equal to zero. Tests for just the

the p slope coefficients are obtained by eliminating $\hat{\beta}_0$ from $\hat{\beta}$ and the relevant row (first) and column (first) from $(\mathbf{X'VX})$. Since evaluation of this test requires the capability to perform vector-matrix operations and to obtain $\hat{\beta}$, there is no gain over the likelihood ratio test. Extensions of the Wald test which might be used to examine functions of the coefficients may be more useful. The modeling approach of Grizzle, Starmer, and Koch (1969), noted earlier, contains many such examples.

The multivariate analog of the Score test is based on the conditional distribution of the p derivatives of $L(\beta)$ with respect to β. The computation of this test is of the same order of complication as the Wald test. To define it in detail would require introduction of additional notation which would find little use in the remainder of this text. Thus, we refer the interested reader to Cox and Hinkley (1974) or Dobson (1983).

2.5 Other Methods of Estimation

In Section 1.4 two alternative methods of estimating the parameters of the logistic regression model were discussed. These were the methods of noniteratively weighted least squares and discriminant function. Each may also be employed in the multivariate case, though application of the noniteratively weighted least squares estimators is limited by the need for nonzero estimates of $\pi(\mathbf{x})$ for most values of $\mathbf{x}$ in the data set. With a large number of independent variables, or even a few continuous variables, this condition is not likely to hold. The discriminant function estimators do not have this limitation and may be easily extended to the multivariate case.

The discriminant function approach to estimation of the logistic coefficients is based on the assumption that the distribution of the independent variables, given the value of the outcome variable, is multivariate normal. Two points should be kept in mind: (1) the assumption of multivariate normality will rarely if ever be satisfied because of the frequent occurrence of dichotomous independent variables, and (2) the discriminant function estimators of the coefficients for nonnormally distributed independent variables, especially dichotomous variables, will be biased away from zero when the true coefficient is nonzero. For these reasons we, in general, do not recommend its use.

However, these estimators are of some historical importance as a number of the classic papers in the applied literature, such as Truett, Cornfield, and Kannel (1967), have used them. These estimators are easily computed and, in the absence of a logistic regression program, should be adequate for a preliminary examination of your data. Thus, it seems worthwhile to include the relevant formulae for their computation.

The assumptions necessary to employ the discriminant function approach to estimation of the logistic regression coefficients state that the conditional distribution of X (the vactor of p covariate random variables) given the outcome variable, $Y = y$, is multivariate normal with a mean vector that depends on y, but a covariance matrix that does not. Using notation defined in Section 1.4 we say $X \mid y = j \sim N(\mu_j, \Sigma)$ where μ_j contains the means of the p independent variables for the subpopulation defined by $y = j$ and Σ is the $p \times p$ covariance matrix of these variables. Under these assumptions $P(Y = 1 \mid x) = \pi(x)$, where the coefficients are given by the following equations:

$$\beta_0 = \ln\left(\frac{\theta_1}{\theta_0}\right) - 0.5(\mu_1 - \mu_0)'\Sigma^{-1}(\mu_1 + \mu_0) \tag{2.6}$$

and

$$\text{ß} = (\mu_1 - \mu_0)'\Sigma^{-1} \tag{2.7}$$

where $\theta_1 = P(Y = 1)$ and $\theta_0 = 1 - \theta_1$, the proportion of the population with y equal to 1 or 0, respectively. Equations (2.6) and (2.7) are the multivariate analogs of equations (1.14) and (1.15).

The discriminant function estimators of β_0 and ß are found by substituting estimators for μ_j, $j = 0, 1$, Σ, and θ_1 into equations (2.6) and (2.7). The estimators most often used are the maximum likelihood estimators under the multivariate normal model. That is, we let

$$\hat{\mu}_j = \bar{x}_j$$

the mean of x in the subgroup of the sample with $y = j$, $j = 0, 1$.

The estimator of the covariance matrix, Σ, is the multivariate extension of the pooled sample variance given in Section 1.4. This may be represented as

$$S = \frac{(n_0 - 1)S_0 + (n_1 - 1)S_1}{(n_0 + n_1 - 2)}$$

where S_j is the $p \times p$ matrix of the usual unbiased estimators of the variances and covariances computed within the subgroup defined by $y = j$, $j = 0, 1$.

Because of the bias in the discriminant function estimators, when normality does not hold, they should be used only when logistic regression software is not available, and then only in preliminary analyses. Any final analyses should be based on the maximum likelihood estimators of the coefficients.

Exercises

1. Use the ICU data described in Section 1.5 and consider the multiple logistic regression model of vital status, STA, on age (AGE), cancer part of the present problem (CAN), CPR prior to ICU admission (CPR), infection probable at ICU admission (INF), and race (RAC).

    1.1 The variable RAC is coded at three levels. Prepare a table showing the coding of the two design variables necessary for including this variable in a logistic regression model.

    1.2 Write down the equation for the logistic regression model of STA on AGE, CAN, CPR, INF, and RAC. Write down the equation for the logit transformation of this logistic regression model. How many parameters does this model contain?

    1.3 Write down an expressions for the likelihood and log-likelihood for the logistic regression model in problem 1.2. How many likelihood equations are there? Write down an expression for a typical likelihood equation for this problem.

    1.4 Using a logistic regression package obtain the maximum likelihood estimates of the parameters of the logistic regression model in problem 1.2. Using these estimates write down the equation for the fitted values, that is, the estimated logistic probabilities.

1.5   Using the results of the output from the logistic regression package used in problem 1.4 assess the significance of the slope coefficients for the variables in the model using the likelihood ratio test. What assumptions are needed for the p-values computed for this test to be valid? What is the value of the deviance for the fitted model?

1.6   Use the Wald statistics to obtain an approximation to the significance of the individual slope coefficients for the variables in the model. Fit a reduced model that eliminates those variables with nonsignificant Wald statistics. Assess the joint (conditional) significance of the variables excluded from the model. Present the results of fitting the reduced model in a table.

1.7   In later chapters we will need the estimated covariance matrix for the estimated parameters in a logistic regression model. Obtain the estimated covariance matrix from the logistic regression package used in problem 1.6 for the model containing AGE, CAN, CPR, INF, and LOC and write it in lower triangular form. If possible, also obtain the estimated information matrix. What is the relationship between these two matrices?

CHAPTER 3

Interpretation of the Coefficients of the Logistic Regression Model

3.1 Introduction

In Chapters 1 and 2 we discussed the methods for fitting and testing for significance of the logistic regression model. After fitting a model the emphasis shifts from the computation and assessment of significance of estimated coefficients to interpretation of their values. Strictly speaking, an assessment of the adequacy of the fitted model should precede any attempt at interpreting it. In the case of logistic regression the methods for assessment of fit are rather technical in nature and thus are deferred until Chapter 5, at which time the reader should have a good working knowledge of the logistic regression model. Thus, we begin this chapter assuming that a logistic regression model has been fit, that the variables in the model are significant in either a biological or statistical sense, and that the model fits according to some statistical measure of fit.

The interpretation of any fitted model requires that we be able to draw practical inferences from the estimated coefficients in the model. The question being addressed is: *What do the estimated coefficients in the model tell us about the research questions that motivated the study?* For most models this will involve the estimated coefficients for the independent variables in the model. On occasion, the intercept coefficient will be of interest; but this is the exception, not the rule. The estimated coefficients for the independent variables represent the slope or rate of change of a function of the dependent variable per unit of change in the independent variable. Thus, interpretation involves two issues: Determining the functional relationship between the dependent variable and the independent variable, and appropriately defining the unit of change for the independent variable.

The first step is to determine what function of the dependent variable yields a linear function of the independent variables. For those with some familiarity with generalized linear models this is called the **link function** [see McCullagh and Nelder (1983)]. In the case of a linear regression model it is the identity function since the dependent variable, by definition, is linear in the parameters. (For those unfamiliar with the term "identity function," it is the function $y = y$.) In the logistic regression model the link function is the logit transformation $g(x) = \ln\{\pi(x)/[1 - \pi(x)]\} = \beta_0 + \beta_1 x$.

For a linear regression model we recall that the slope coefficient, β_1, is equal to the difference between the value of the dependent variable at $x + 1$ and the value of the dependent variable at x, for any value of x. To demonstrate this, let $y(x) = \beta_0 + \beta_1 x$ from which it follows that $\beta_1 = y(x + 1) - y(x)$. In this case, the interpretation of the coefficient is relatively straightforward, as it expresses the resulting change in the measurement scale of the dependent variable for a unit change in the independent variable. For example, if in a regression of weight on height of adolescent aged male children we found the slope to be 5, then we would conclude that a change of 1 inch in height is associated with a change of 5 pounds in weight.

In the logistic regression model $\beta_1 = g(x + 1) - g(x)$. That is, the slope coefficient represents the change in the logit for a change of one unit in the independent variable x. Proper interpretation of the coefficient in a logistic regression model depends on being able to place meaning on the difference between two logits. Interpretation of this difference will be discussed in detail on a case-by-case basis as it relates directly to the definition and meaning of a one unit change in the independent variable. In the following sections of this chapter we will consider the interpretation of the coefficients for a univariate logistic regression model for each of the possible measurement scales of the independent variable.

3.2 Dichotomous Independent Variable

We begin our consideration of the interpretation of logistic regression coefficients with the situation where the independent variable is dichotomous. This case is not only the simplest, but it will provide the conceptual foundation for all the other situations. We assume that x is coded as either zero or 1. Under this model there are two values of $\pi(x)$ and equivalently two values of $1 - \pi(x)$. These values may be conveniently displayed in a 2×2 table as shown in Table 3.1.

Table 3.1   Values of the Logistic Regression Model When the Independent Variable Is Dichotomous.

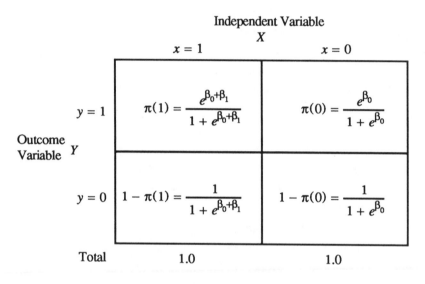

		Independent Variable X	
		$x = 1$	$x = 0$
Outcome Variable Y	$y = 1$	$\pi(1) = \dfrac{e^{\beta_0+\beta_1}}{1 + e^{\beta_0+\beta_1}}$	$\pi(0) = \dfrac{e^{\beta_0}}{1 + e^{\beta_0}}$
	$y = 0$	$1 - \pi(1) = \dfrac{1}{1 + e^{\beta_0+\beta_1}}$	$1 - \pi(0) = \dfrac{1}{1 + e^{\beta_0}}$
	Total	1.0	1.0

The **odds** of the outcome being present among individuals with $x = 1$ is defined as $\pi(1)/[1 - \pi(1)]$. Similarly, the odds of the outcome being present among individuals with $x = 0$ is defined as $\pi(0)/[1 - \pi(0)]$. The log of the odds, as defined previously, is called the logit and, in this example, these are

$$g(1) = \ln\{\pi(1)/[1 - \pi(1)]\}$$

and

$$g(0) = \ln\{\pi(0)/[1 - \pi(0)]\}$$

The **odds ratio**, denoted by ψ, is defined as the ratio of the odds for $x = 1$ to the odds for $x = 0$, and is given by the equation

$$\psi = \frac{\pi(1)/[1 - \pi(1)]}{\pi(0)/[1 - \pi(0)]} \tag{3.1}$$

The log of the odds ratio, termed log-odds ratio, or **log-odds**, is

$$\ln(\psi) = \ln\left[\frac{\pi(1)/[1 - \pi(1)]}{\pi(0)/[1 - \pi(0)]}\right]$$

$$= g(1) - g(0)$$

which is the **logit difference**.

Now, using the expressions for the logistic regression model shown in Table 3.1 the odds ratio is

$$\psi = \frac{\left(\dfrac{e^{\beta_0 + \beta_1}}{1 + e^{\beta_0 + \beta_1}}\right)\left(\dfrac{1}{1 + e^{\beta_0}}\right)}{\left(\dfrac{e^{\beta_0}}{1 + e^{\beta_0}}\right)\left(\dfrac{1}{1 + e^{\beta_0 + \beta_1}}\right)}$$

$$= \frac{e^{\beta_0 + \beta_1}}{e^{\beta_0}} = e^{\beta_1}$$

Hence, for logistic regression with a dichotomous independent variable

$$\psi = e^{\beta_1} \tag{3.2}$$

and the logit difference, or log odds, is

$$\ln(\psi) = \ln(e^{\beta_1}) = \beta_1$$

This fact concerning the interpretability of the coefficients is the fundamental reason why logistic regression has proven such a powerful analytic tool for epidemiologic research.

The odds ratio is a measure of association which has found wide use, especially in epidemiology, as it approximates how much more likely (or unlikely) it is for the outcome to be present among those with $x = 1$ than among those with $x = 0$. For example, if y denotes the presence or absence of lung cancer and if x denotes whether or not the person is a smoker, then $\hat{\psi} = 2$ indicates that lung cancer occurs twice as often among smokers than among nonsmokers in the study population. As another example, suppose y denotes the presence or absence of CHD and x denotes whether or not the person engages in regular strenuous physical exercise. Then if $\psi = 0.5$, the occurrence of CHD is one half as frequent among those who exercise than among those who do not in the study population.

Readers who have not had experience with the odds ratio as a measure of association would be advised to spend some time reading about this measure in one of the following texts: Breslow and Day (1980), Kelsey, Thompson, and Evans (1986), Rothman (1986), and Schlesselman (1982).

The interpretation given for the odds ratio is based on the fact that in many instances it approximates a quantity called the relative risk. This parameter will be represented in this chapter as η and is equal to the ratio $\pi(1)/\pi(0)$. It follows from equation (3.1) that $\psi \approx \eta$ if $[1 - \pi(0)/[1 - \pi(1)] \approx 1$. This approximation will hold when $\pi(x)$ is small for both $x = 1$ and 0.

An example may help to clarify what the odds ratio is and how it is computed from a logistic regression program or from a 2×2 table. In many examples of logistic regression encountered in the literature we find that a continuous variable has been dichotomized at some biologically meaningful cutpoint. A more detailed discussion of the rationale and implications for modeling of such a decision will be presented in Chapter 4. With this in mind we will use the data displayed in Table 1.1 and create a new variable which takes on the value 1 if the age of the subject is greater than or equal to 55 and zero otherwise. The result of cross-classifying the dichotomized age variable with the outcome variable CHD is presented in Table 3.2.

Table 3.2 Cross-Classification of AGE Dichotomized at 55 Years and CHD for 100 Subjects.

CHD (y)	AGE (x)		Total
	55 (1)	<55 (0)	
Present (1)	21	22	43
Absent (0)	6	51	57
Total	27	73	100

The data in Table 3.2 tell us that there were 21 subjects with values $(x = 1, y = 1)$, 22 with $(x = 0, y = 1)$, 6 with $(x = 0, y = 0)$, and 51 with $(x = 0, y = 0)$. Hence, evaluating the likelihood function shown in equation (1.3) for these data is

$$l(\beta) = \pi(1)^{21} \times [1 - \pi(1)]^{6} \times \pi(0)^{22} \times [1 - \pi(0)]^{51}$$

Use of a logistic regression program to find the values of β_0 and β_1 which maximize this likelihood yields the following estimated coefficients and standard errors.

Table 3.3 Results of Fitting the Logistic Regression Model to the Data in Table 3.2.

Variable	Estimated Coefficient	Standard Error	Coeff./SE	$\hat{\psi}$
AGE	2.094	0.529	3.96	8.1
Constant	−0.841	0.255	−3.30	

The quantity in the column labeled $\hat{\psi}$ in Table 3.3 is the maximum likelihood estimate of the odds ratio, $\hat{\psi} = e^{2.094} = 8.1$. Readers who have had some previous experience with the odds ratio will undoubtedly wonder why a logistic regression package was used to obtain the maximum likelihood estimate of the odds ratio, when it could have been obtained directly from the cross-product ratio from Table 3.2., namely

$$\hat{\psi} = \frac{21/6}{22/51} = 8.11$$

and $\hat{\beta}_1 = \ln[(21/6)/(22/51)] = 2.094$. The reason for this is to emphasize that logistic regression is, in fact, regression even in the simplest case possible. The fact that the data may be formulated in terms of a contingency table will provide us with the basis for interpretation of estimated coefficients as the log of odds ratios.

Along with the point estimate of a parameter, a confidence interval estimate may also be employed to provide additional information. In the case of the odds ratio, ψ, for a 2×2 table there is an extensive literature dealing with this problem, much of which is focused on methods when the sample size is small. The reader who wishes to learn more about the available exact and approximate methods should see the papers by Fleiss (1979) and Gart and Thomas (1972). A good summary may be found in the texts by Breslow and Day (1980), Kleinbaum, Kupper, and Morgenstern (1982), and Rothman (1986). Software implementing exact methods is available in the EGRET program and will be available in SYSTAT.

The odds ratio, ψ, is usually the parameter of interest in a logistic regression due to its ease of interpretation. However, its estimate, $\hat{\psi}$, will tend to have a distribution that is skewed. The skewness of the sampling distribution of $\hat{\psi}$ is due to the fact that it is bounded away from zero. In theory, for large enough sample sizes, the distribution of $\hat{\psi}$ will be normal. Unfortunately, this sample size requirement usually exceeds that of most studies. Hence, inferences are usually based on the sampling distribution of $\ln(\hat{\psi}) = \hat{\beta}_1$, which tends to follow a normal distribution for much smaller sample sizes. A $100 \times (1 - \alpha)\%$ confidence interval (CI) estimate for the odds ratio is obtained by first calculating the endpoints of a confidence interval for the coefficient β_1, and then exponentiating these values. In general, the endpoints are given by the expression

$$\exp[\hat{\beta}_1 \pm z_{1-\alpha/2} \times \hat{SE}(\hat{\beta}_1)]$$

Because of the importance of the odds ratio as a measure of association, point and interval estimates are often found in additional columns in tables presenting the results of a logistic regression analysis.

As an example consider the estimation of the odds ratio for the dichotomized variable age. The point estimate is $\hat{\psi} = 8.1$ and the endpoints of a 95% CI are

$$\exp(2.094 \pm 1.96 \times 0.529) = (2.9, 22.9)$$

This interval is typical of confidence intervals for the odds ratio when the point estimate exceeds 1. The interval is skewed to the right.

Before concluding the dichotomous variable case some mention of the effect of coding and the role of design variables is in order. In the previous discussion we noted that the estimate of the odds ratio was $\hat{\psi} = \exp(\hat{\beta}_1)$. This is correct when the independent variable has been coded as zero or 1. Other coding may require that we calculate the value of the logit difference for the specific coding used, and then exponentiate.

It will be helpful to illustrate these computations in detail, as they demonstrate the general method for computation of estimates of odds ratios in logistic regression. The estimate of the log odds (log of the odds ratio) for any independent variable at two different levels, say $x = a$ versus $x = b$, is the difference between the estimated logits computed at these two values. The equations that specify this calculation are as follows:

$$\ln[\hat{\psi}(a,b)] = \hat{g}(x = a) - \hat{g}(x = b)$$
$$= (\hat{\beta}_0 + \hat{\beta}_1 \times a) - (\hat{\beta}_0 + \hat{\beta}_1 \times b)$$
$$= \hat{\beta}_1 \times (a - b) \tag{3.3}$$

and the estimated odds ratio is

$$\hat{\psi}(a,b) = \exp[\hat{\beta}_1 \times (a - b)] \tag{3.4}$$

This expression is equal to $\exp(\hat{\beta}_1)$ only when $(a - b) = 1$. In equations (3.3) and (3.4) the notation $\hat{\psi}(a,b)$ is used to represent the odds ratio

$$\hat{\psi}(a,b) = \frac{\hat{\pi}(x = a)/[1 - \hat{\pi}(x = a)]}{\hat{\pi}(x = b)/[1 - \hat{\pi}(x = b)]} \tag{3.5}$$

and when $a = 1$ and $b = 0$ we let $\hat{\psi} = \hat{\psi}(1,0)$.

When we are using a dichotomous independent variable we have a choice of whether or not to use a design variable. A dichotomous variable will require a single design variable with a single estimated coefficient, while treating the dichotomous variable as interval scaled will yield an equivalent, though not necessarily numerically equal, coefficient.

The logistic regression program BMDPLR in BMDP offers two choices for forming design variables when the independent variable is dichotomous. These are called the **marginal method** and **partial method**. The marginal method is also referred to as the **deviation from means** parameterization and the partial method is also referred to as the **reference cell** parameterization. The partial method assigns the value of zero to the lowest code for x and one to the highest. If, for example, SEX was coded as 1 = male and 3 = female, then the resulting design variable under this method, D, would be coded 0 = male and 1 = female. Exponentiation of the estimated coefficient for D would estimate the odds ratio of female relative to male. This same result would have been obtained had sex been coded originally as 0 = male and 1 = female, and then treating the variable SEX as if it were interval scaled.

The second method of coding design variables in BMDPLR is called the marginal method. This method assigns the value of −1 to the lower code, and a value of 1 to the higher code. The coding for the variable SEX discussed above is shown in Table 3.4.

Table 3.4　Illustration of the Coding of the Design Variable Using the Marginal Method.

SEX (Code)	Design Variable D
Male (1)	−1
Female (3)	1

Suppose we wish to estimate the odds ratio of female versus male when the marginal method of coding is used. We do this by using the general method shown in equations (3.3) and (3.4),

$$\ln[\hat{\psi}(\text{female},\text{male})] = \hat{g}(\text{female}) - \hat{g}(\text{male})$$
$$= \hat{g}(D = 1) - \hat{g}(D = -1)$$
$$= [\hat{\beta}_0 + \hat{\beta}_1 \times (D = 1)] - [\hat{\beta}_0 + \hat{\beta}_1 \times (D = -1)]$$
$$= 2\,\hat{\beta}_1$$

and the estimated odds ratio is $\hat{\psi}(\text{female},\text{male}) = \exp(2\hat{\beta}_1)$. Thus, if we had blindly exponentiated the coefficient from the computer output we would have obtained the wrong estimate of the odds ratio. This points out quite clearly that we must pay close attention to the method used to code the design variables.

The method of coding also influences the calculation of the endpoints of the confidence interval. For the above example, using the marginal method, the estimated standard error needed for confidence interval estimation is $\hat{SE}(2\hat{\beta}_1)$ which is $2 \times \hat{SE}(\hat{\beta}_1)$. Thus the endpoints of the confidence interval are

$$\exp[2\hat{\beta}_1 \pm z_{1-\alpha/2}2\hat{SE}(\hat{\beta}_1)]$$

In general, the endpoints of the confidence interval for the odds ratio given in equation (3.5) are $\exp[\hat{\beta}_1(a-b) \pm z_{1-\alpha/2}|a-b| \times \hat{SE}(\hat{\beta}_1)]$, where $|a-b|$ is the absolute value of $(a-b)$. Since we have control over how we code our dichotomous variables, use of a design variable to represent a dichotomous variable is redundant. We recommend that all dichotomous variables be coded, or recoded for analysis, as zero or 1, and that each be treated as interval scaled.

In summary, for a dichotomous variable the parameter of interest is the odds ratio. An estimate of this parameter may be obtained from the estimated logistic regression coefficient, regardless of how the variable is coded or scaled. This relationship between the logistic regression coefficient and the odds ratio provides the foundation for our interpretation of all logistic regression results.

3.3 Polytomous Independent Variable

Suppose that instead of two categories the independent variable has $k > 2$ distinct values. For example, we may have variables that denote the county of residence within a state, the clinic used for primary health care within a city, or race. Each of these variables has a fixed number of discrete outcomes and the scale of measurement is nominal. We saw in Chapter 2 that it is inappropriate to model a nominal scaled variable as if it were interval scaled. Therefore, we must form a set of design variables to represent the categories of the variable. In this section we will present methods for creating design variables for polytomous independent variables. The choice of a particular method will depend to some extent on the goals of the analysis and the stage of model development.

We begin by extending the method that was presented in Table 2.1 for a dichotomous variable. For example, suppose that in a study of CHD the variable RACE is coded at four levels, and that the cross-classification of RACE by CHD status yields the data in Table 3.5. These data are hypothetical and have been formulated for ease of computation. The extension to a situation where the variable has more than four levels is not conceptually different, so all the examples in this section will use $k = 4$.

Table 3.5 Cross-Classification of Hypothetical Data on RACE and CHD Status for 100 Subjects.

CHD Status	White	Black	Hispanic	Other	Total
Present	5	20	15	10	50
Absent	20	10	10	10	50
Total	25	30	25	20	100
Odds Ratio ($\hat{\psi}$)	1.0	8.0	6.0	4.0	
95% CI		(2.3,27.6)	(1.7,21.3)	(1.1,14.9)	
$\ln(\hat{\psi})$	0.0	2.08	1.79	1.39	

At the bottom of Table 3.5 the odds ratio is given for each race, using white as the reference group. For example, for hispanic the estimated odds ratio is $(15 \times 20)/(5 \times 10) = 6.0$. The log of the odds ratios are given in the last row of Table 3.5. This display is typical of what is found in the literature when there is a perceived referent group to which the other groups are to be compared. These same estimates of the odds ratio may be obtained from a logistic regression program with an appropriate choice of design variables. The method for specifying the design variables involves setting all of them equal to zero for the reference group, and then setting a single design variable equal to 1 for each of the other groups. This is illustrated in Table 3.6.

This method, sometimes referred to as "reference cell coding," is used in the program BMDPLR where it is called the partial method. It is also the method used in the programs EGRET and GLIM. SAS and SYSTAT use the same approach, but the group with the largest code serves as the reference group, the category Other in the example. To obtain the same set of design variables in

SAS or SYSTAT as is used in EGRET, GLIM, and BMDP we would have to reverse the coding to 1 = black, 2 = hispanic, 3 = other, and 4 = white.

Table 3.6 Specification of the Design Variables for RACE Using White as the Reference Group.

RACE (Code)	Design Variables		
	D_1	D_2	D_3
White (1)	0	0	0
Black (2)	1	0	0
Hispanic (3)	0	1	0
Other (4)	0	0	1

Use of any logistic regression program with design variables coded as shown in Table 3.6 yields the estimated logistic regression coefficients given in Table 3.7.

Table 3.7 Results of Fitting the Logistic Regression Model to the Data in Table 3.5 Using the Design Variables in Table 3.6.

Variable	Estimated Coefficient	Standard Error	Coeff./SE	$\hat{\psi}$
RACE (1)	2.079	0.633	3.29	8.0
RACE (2)	1.792	0.646	2.78	6.0
RACE (3)	1.386	0.671	2.07	4.0
Constant	−1.386	0.500	−2.77	

A comparison of the estimated coefficients in Table 3.7 to the log odds in Table 3.5 shows that $\ln[\hat{\psi}(\text{black,white})] = \hat{\beta}_{11} = 2.079$, $\ln[\hat{\psi}(\text{hispanic,white})] = \hat{\beta}_{12} = 1.792$ and $\ln[\hat{\psi}(\text{other,white})] = \hat{\beta}_{13} = 1.386$. Did this happen by chance? Calculation of the logit difference shows that it is by design. For the comparison of black to white this is as follows:

$$\ln[\hat{\psi}(\text{black,white})] = \hat{g}(\text{black}) - \hat{g}(\text{white})$$

$$= [\hat{\beta}_0 + \hat{\beta}_{11} \times (D_1 = 1) + \hat{\beta}_{12} \times (D_2 = 0) + \hat{\beta}_{13} \times (D_3 = 0)]$$

$$- [\hat{\beta}_0 + \hat{\beta}_{11} \times (D_1 = 0) + \hat{\beta}_{12} \times (D_2 = 0) + \hat{\beta}_{13} \times (D_3 = 0)]$$

$$= \hat{\beta}_{11}$$

Calculation of other comparisons would demonstrate equality between other estimated coefficients and the log of odds ratios computed from the data in Table 3.5.

A comment about the estimated standard errors may be helpful at this point. In the univariate case the estimates of the standard errors found in the logistic regression output are identical to the estimates obtained using the cell frequencies from the contingency table. For example, the estimated standard error of the estimated coefficient for design variable (1), $\hat{\beta}_{11}$, is $0.6325 = (1/5 + 1/20 + 1/20 + 1/10)^{1/2}$. A derivation of this result may be found in Bishop, Feinberg, and Holland (1975).

Confidence limits for odds ratios may be obtained using the same approach used in Section 3.2 for a dichotomous variable. We begin by finding the confidence limits for the log odds (the logistic regression coefficient) and then exponentiating these limits to obtain limits for the odds ratio. In general, the limits for a $100(1 - \alpha)\%$ CI for the coefficient are of the form

$$\hat{\beta}_{ij} \pm z_{1-\alpha/2} \times \hat{SE}(\hat{\beta}_{ij})$$

The corresponding limits for the odds ratio are obtained by exponentiating these limits as follows:

$$\exp[\hat{\beta}_{ij} \pm z_{1-\alpha/2} \times \hat{SE}(\hat{\beta}_{ij})] \tag{3.6}$$

The confidence limits given in Table 3.5 in the row beneath the estimated odds ratios were obtained using equation (3.6) for $j = 1, 2, 3$ with $\alpha = 0.05$.

The method of coding design variables which uses a referent group is the most commonly employed method appearing in the literature. The primary reason for the widespread use of this method is the interest in estimating the risk of an "exposed" group relative to that of a "control" or "unexposed" group.

A second method of coding design variables, which is more frequently used in analysis of variance and linear regression than in logistic regression, is called "deviation from means coding." This expresses effect as the deviation of the "group mean" from the "overall mean." In the case of logistic regression the "group mean" is the logit for the group and the "overall mean" is the average logit. The required coding is obtained, as illustrated for race in Table 3.8, by setting the value of all the design variables equal to −1 for one of the categories, and then using the 0,1 coding for the remainder of the categories.

Table 3.8 Specification of the Design Variables for RACE to Yield the Deviation from the Average Logit.

	Design Variables		
RACE (Code)	D_1	D_2	D_3
White (1)	−1	−1	−1
Black (2)	1	0	0
Hispanic (3)	0	1	0
Other (4)	0	0	1

Use of the design variables in Table 3.8 yields the estimated logistic regression coefficients given in Table 3.9.

Table 3.9 Results of Fitting the Logistic Regression Model to the Data in Table 3.5 Using the Design Variables in Table 3.8.

Variable	Estimated Coefficient	Standard Error	Coeff./SE
RACE (1)	0.765	0.351	2.18
RACE (2)	0.477	0.362	1.32
RACE (3)	0.719E-01	0.385	0.19
Constant	−0.719E-01	0.500	−0.33

In order to interpret the estimated coefficients in Table 3.9 we need to refer to Table 3.5 and calculate the logit for each of the four categories of RACE. These are

$$\hat{g}_1 = \ln\left(\frac{5/25}{20/25}\right) = \ln\left(\frac{5}{20}\right) = -1.386$$

$\hat{g}_2 = \ln(20/10) = 0.693$, $\hat{g}_3 = \ln(15/10) = 0.405$, $\hat{g}_4 = \ln(10/10) = 0$, and their average is $\bar{g} = \Sigma\hat{g}_j/4 = -0.072$ The estimated coefficient for design variable (1) in Table 3.9 is $0.765 = \hat{g}_2 - \bar{g} = 0.693 - (-0.072)$. The general relationship for the estimated coefficient for design variable (j) is $\hat{g}_{j+1} - \bar{g}$, for $j = 1, 2, 3$.

The interpretation of the estimated coefficients is not as easy or clear as in the situation when a referent group is used. Exponentiation of the estimated coefficients yields the ratio of the odds for the particular group to the geometric mean of the odds. Specifically, for design variable (1) in Table 3.9 we have

$$\exp(0.765) = \exp(\hat{g}_2 - \bar{g})$$
$$= \exp(\hat{g}_2)/\exp(\Sigma\,\hat{g}_j/4)$$
$$= (20/10)/[(5/20)\times(20/10)\times(15/10)\times(10/10)]^{0.25}$$
$$= 2.15$$

This number, 2.15, is not a true odds ratio because the quantities in the numerator and denominator do not represent the odds for two distinct categories. The exponentiation of the estimated coefficient expresses the odds relative to an "average" odds, the geometric mean. Whether this is in fact useful will depend on being able to place a meaningful interpretation on the "average" odds.

The estimated coefficients obtained from the deviation from means coding may be used to estimate the odds ratio for one category relative to a referent category. However, the equation for the estimate is more complicated than the one obtained using the reference cell coding.

To illustrate this consider estimation of the log odds of black versus white using the coding for design variables given in Table 3.8. The logit difference is as follows:

$$\ln[\hat{\psi}(\text{black},\text{white})] = \hat{g}(\text{black}) - \hat{g}(\text{white})$$
$$= [\hat{\beta}_0 + \hat{\beta}_{11}\times(D_1 = 1) + \hat{\beta}_{12}\times(D_2 = 0) + \hat{\beta}_{13}\times(D_3 = 0)]$$
$$- [\hat{\beta}_0 + \hat{\beta}_{11}\times(D_1 = -1) + \hat{\beta}_{12}\times(D_2 = -1) + \hat{\beta}_{13}\times(D_3 = -1)]$$
$$= 2\hat{\beta}_{11} + \hat{\beta}_{12} + \hat{\beta}_{13} \qquad\qquad (3.7)$$

To obtain a confidence interval we must estimate the variance of the sum of the coefficients given in equation (3.7). In this case it is as follows:

$$\text{var}\{\ln[\hat{\psi}(\text{black},\text{white})]\} = 4 \text{ var}(\hat{\beta}_{11}) + \text{var}(\hat{\beta}_{12}) + \text{var}(\hat{\beta}_{13})$$

$$+ 4 \text{ cov}(\hat{\beta}_{11},\hat{\beta}_{12}) + 4 \text{ cov}(\hat{\beta}_{11},\hat{\beta}_{13}) + 2 \text{ cov}(\hat{\beta}_{12},\hat{\beta}_{13}) \qquad (3.8)$$

Each of the terms in the above expression may be obtained from output that is available from most logistic regression software. Confidence intervals for the odds ratio are obtained by exponentiating the endpoints of the confidence limits for the sum of the coefficients in equation (3.7). Evaluation of equation (3.7) for the current example gives

$$\ln[\hat{\psi}(\text{black},\text{white})] = 2(0.765) + 0.477 + 0.072 = 2.079$$

An estimate of the variance is obtained by evaluating equation (3.8) which, for the current example, yields

$$\hat{\text{var}}\{\ln[\hat{\psi}(\text{black},\text{white})]\} = 4(0.351)^2 + (0.362)^2 + (0.385)^2$$

$$+ 4(-0.031) + 4(-0.040) + 2(-0.044) = 0.400$$

and

$$\hat{\text{SE}}\{\ln[\hat{\psi}(\text{black},\text{white})]\} = 0.633$$

We note that the values of the estimated log odds, 2.079, and the estimated standard error, 0.633, are identical to the values of the estimated coefficient and standard error for the first design variable in Table 3.7. This should be expected, since the design variables used to obtain the estimated coefficients in Table 3.7 were formulated specifically to yield the log odds relative to the white race category.

It should be apparent that use of deviation from average logit design variables is computationally much more complex than the referent cell method for estimation of odds ratios. A more detailed and general treatment of estimation of odds ratios with this type of design variables may be found in Lemeshow and Hosmer (1983).

There is one additional method of creating design variables which may be useful when the categorical variable is at least ordinal scaled. This method is based on orthogonal polynomials. Readers not familiar with the use of orthogonal polynomials in linear regression and analysis of variance may wish to refer to Fleiss (1986), or to skip the remainder of this section.

Orthogonal polynomials are typically used to assess the trend in the response variable over increasing levels of an interval-scaled independent variable. Therefore, in order to present the method of orthogonal polynomials in the correct context we will use the data from Example 1.1. The example data are obtained by collapsing the eight groups in Table 1.2 into four age groups by pooling groups 1 and 2, 3 and 4, and so forth. The resulting contingency table is given in Table 3.10.

Table 3.10 Cross-Classification of Age Group and CHD Status for 100 Subjects Shown in Table 1.1.

CHD Status	Age Group				Total
	20–34	35–44	45–54	55–64	
Present	3	8	11	21	43
Absent	22	19	10	6	57
Total	25	27	21	27	100
Odds Ratio($\hat{\psi}$)	1.0	3.1	8.1	25.7	
$\ln(\hat{\psi})$	0.0	1.1	2.1	3.2	

The log odds ratios show that relative to the referent group the risk of CHD increases with age. A natural question to ask is if this increase is systematically related to age. An empirical assessment may be obtained by plotting the log odds ratios versus the midpoint of the age groups. This is illustrated in Chapter 4. An alternative approach is possible in this case since the midpoints of the age groups are nearly equally spaced (27.5, 40, 50, 60). We may create three design variables using coefficients for orthogonal polynomials. The coding shown in Table 3.11 is that provided by BMDPLR for this example.

Table 3.11 Specification of the Design Variables for Age Group to Yield the Orthogonal Polynomials.

Age Group	Design Variables		
	D_1	D_2	D_3
20–34	−0.6708	0.5	−0.2236
35–44	−0.2236	−0.5	0.6708
45–54	0.2236	−0.5	−0.6708
55–64	0.6708	0.5	0.2236

The coefficients shown in this table differ by a multiplicative constant from the coefficients for orthogonal polynomials in analysis of variance and regression [See Fleiss (1986), page 229]. The resulting estimated logistic regression coefficients provide us with the means to test whether the relationship between age and the logit has significant linear, quadratic, and/or cubic components. The estimated coefficients resulting from this coding are contained in Table 3.12.

Table 3.12 Results of Fitting the Logistic Regression Model to the Data in Table 3.10 Using the Design Variables in Table 3.11.

Variable	Estimated Coefficient	Standard Error	Coeff./SE
Age Group (1)	2.382	0.534	4.48
Age Group (2)	0.150E-01	0.490	0.03
Age Group (3)	0.814E-01	0.442	0.18
Constant	−0.377E-01	0.245	−0.54

Of interest in this case are the univariate Wald statistics, the ratio of the estimated coefficient to its estimated standard error. These Wald statistics may be used to test for the significance of the components of trend. In the current example only the Wald statistic for the first design variable is significant, which leads us to conclude that there is strong evidence for the logit to increase linearly in age. The result of this analysis is that in future modeling we may choose to use age as a continuous variable, and model it with a single parameter rather than with the three parameters required when we use one of the methods for design variables.

The orthogonal method of creating design variables may be obtained in BMDPLR using the orthogonal option. They may be obtained from other programs as well, but require the user to specify the coding.

In summary, we have shown that discrete nominal scaled variables are included properly into the analysis only when they have been recoded into design variables. The particular choice of design variables will depend on the application, though the referent group type of coding is judged to be the easiest to interpret, and thus is the most commonly used method.

3.4 Continuous Independent Variable

When a logistic regression model contains a continuous independent variable, interpretation of the estimated coefficient will depend on how it is entered into the model and the particular units of the variable. For purposes of developing the method to interpret the coefficient for a continuous variable, we will assume that the logit is linear in the variable. Other modeling strategies that examine this assumption will be presented in Chapter 4.

Under the assumption that the logit is linear in the continuous covariate, x, the equation for the logit is $g(x) = \beta_0 + \beta_1 x$. It follows that the slope coefficient, β_1, gives the change in the log odds for an increase of "1" unit in x, that is, $\beta_1 = g(x + 1) - g(x)$ for any value of x. Most often the value of "1" will not be biologically very interesting. For example, an increase of 1 year in age or of 1 mm Hg in systolic blood pressure may be too small to be considered important. A change of 10 years or 10 mm Hg might be considered more useful. On the other hand, if the range of x is from zero to 1, as might be the case for some created index , then a change of 1 is too large and a change of 0.01 may be more realistic. Hence, to provide a useful interpretation for continuous scaled covariates we need to develop a method for point and interval estimation for an arbitrary change of "c" units in the covariate.

The log odds for a change of c units in x is obtained from the logit difference $g(x + c) - g(x) = c\beta_1$ and the associated odds ratio is obtained by exponentiating this logit difference, $\psi(c) = \psi(x + c, x) = \exp(c\beta_1)$. An estimate may be obtained by replacing β_1 with its maximum likelihood estimate $\hat{\beta}_1$. An estimate of the standard error needed for confidence interval estimation is obtained by multiplying the estimated standard error of $\hat{\beta}_1$ by c. Hence the endpoints of the $100(1 - \alpha)\%$ CI estimate of $\psi(c)$ are

$$\exp[c\hat{\beta}_1 \pm z_{1-\alpha/2}c \; \hat{SE}(\hat{\beta}_1)]$$

Since both the point estimate and endpoints of the confidence interval depend on the choice of c, the particular value of c should be clearly specified in all tables and calculations. The rather arbitrary nature of the choice of c may be somewhat troublesome to some. For example, why use a change of 10 years when 5 or 15 or even 20 years may be equally good? We of course could use any reasonable value; but the goal must be kept in mind: to provide the reader of your analysis with a clear indication of how the risk of the outcome being present changes with the variable in question. Changes in multiples of 5 or 10 may be most meaningful and easily understood.

As an example, consider the univariate model in Table 1.3. In that example a logistic regression of AGE on CHD status using the data of Table 1.1 was reported. The resulting estimated logit was $\hat{g}(AGE) = -5.310 + 0.111 \times AGE$. The estimated odds ratio for an increase of 10 years in age is $\hat{\psi}(10) = \exp(10 \times 0.111) = 3.03$. This indicates that for every increase of 10 years in age, the risk of CHD increases 3.03 times. The validity of such a statement is questionable in this example, since the additional risk of CHD for a 40 year-old compared to a 30 year-old may be quite different from the additional risk of CHD for a 60 year-old compared to a 50 year-old. This is an unavoidable dilemma when continuous covariates are modeled linearly in the logit. If it is believed that the logit is not linear in the covariate, then grouping and use of dummy variables should be considered. Alternatively, use of higher order terms (e.g., $x^2, x^3, ...$) or nonlinear scaling in the covariate (e.g., $\log x$) could be considered. Thus, we see that an important modeling consideration for continuous covariates is their scale in the logit. We consider this in some detail in Chapter 4. The endpoints of a 95% confidence interval for this odds ratio are

$$\exp(10 \times 0.111 \pm 1.96 \times 10 \times 0.024) = (1.90, 4.86)$$

Results similar to these may be placed in tables displaying the results of a fitted logistic regression model.

In summary, the interpretation of the estimated coefficient for a continuous variable is similar to that of nominal scaled variables: an estimated log odds ratio. The primary difference is that a meaningful change must be defined for the continuous variable.

3.5 Multivariate Case

In the previous sections in this chapter we discussed the interpretation of an estimated logistic regression coefficient in the case when there is a single variable in the fitted model. Fitting a series of univariate models will rarely provide an adequate analysis of the data in any study since the independent variables are usually not associated with one another and may have different distributions within levels of the outcome variable. Thus, one generally considers a multivariate analysis for a more comprehensive modeling of the data. One goal of such an analysis is to **statistically adjust** the estimated effects of each variable in the model for differences in the distributions of and associations among the other independent variables. Applying this concept to a multivariate logistic regression model, we may surmise that each estimated coefficient provides an estimate of the log odds adjusting for all other variables included in the model.

To fully understand the estimates of the coefficients from a multivariate logistic regression model requires that we have a clear understanding of what is actually meant by the term *adjusting, statistically, for other variables*. We begin by examining adjustment in the context of a linear regression model, and then extend the concept to logistic regression.

The multivariate situation we will examine is one in which the model contains two independent variables – one dichotomous and one continuous – but primary interest is focused on the effect of the dichotomous variable. This situation is frequently encountered in epidemiologic research when an exposure to a risk factor is recorded as being either present or absent, and we wish to adjust for a variable such as age. The analogous situation in linear regression is called **analysis of covariance**.

Suppose we wish to compare the mean weight of two groups of boys. It is known that weight is associated with many characteristics, one of which is age. Assume that on all characteristics except age the two groups have nearly identical distributions. If the age distribution is also the same for the two groups, then a univariate analysis would suffice and we could compare the mean weight of the two groups. This comparison would provide us with a correct estimate of the difference between the two groups. However, if one group was much younger than the other group, then a comparison of the two groups would be meaningless, since at least a portion of any difference observed would likely be due to the difference in age. It would not be possible to determine the effect of group without first eliminating the discrepancy in ages between the groups.

This situation is described graphically in Figure 3.1. In this figure it is assumed that the relationship between age and weight is linear, with the same nonzero slope in each group. Both of these assumptions would usually be tested in an analysis of covariance before making any inferences about group differences. We will defer a discussion of this until Chapter 4, as it gets to the heart of modeling with logistic regression. We will proceed as if these assumptions have been checked and are supported by the data.

The statistical model that describes the situation in Figure 3.1 states that the value of weight, w, may be expressed as $w = \beta_0 + \beta_1 x + \beta_2 a$, where $x = 0$ for group 1 and $x = 1$ for group 2 and "a" denotes age. In this model the parameter β_1 represents the true difference in weight between the two groups and β_2 is the rate of change in weight per year of age. Suppose that the mean age of group 1 is $\bar{a}_1$ and the mean age of group 2 is $\bar{a}_2$. These values are indicated in Figure 3.1. Comparison of the mean weight of group 1 to the mean weight of group 2 amounts to a comparison of w_1 to w_2. In terms of the model this difference is $(w_2 - w_1) = \beta_1 + \beta_2(\bar{a}_2 - \bar{a}_1)$. Thus the comparison involves not only the true difference between the groups, β_1, but a component, $\beta_2(\bar{a}_2 - \bar{a}_1)$, which reflects the difference between the ages of the groups.

The process of statistically adjusting for age involves comparing the two groups at some common value of age. The value usually used is the mean of the two groups which, for the example, is denoted by $\bar{a}$ in Figure 3.1. In terms of the model this yields a comparison of w_4 to w_3, $(w_4 - w_3) = \beta_1 + \beta_2(\bar{a} - \bar{a}) = \beta_1$, the true difference between the two groups. In theory any common value of age could be used as it would yield the same difference between the two lines. The choice of the overall mean makes sense for two reasons: it is biologically reasonable and lies within the range for which we believe that the association between age and weight is linear and constant within each group.

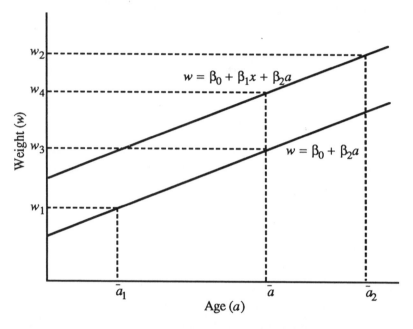

Figure 3.1   Comparison of the Weight of Two Groups of Boys with Different Distributions of Age.

Consider the same situation shown in Figure 3.1, but instead of weight being the dependent variable, let it be a dichotomous variable and let the vertical axis denote the logit. That is, under the model the logit, $g(x,a)$, is given by the equation $g(x,a) = \beta_0 + \beta_1 x + \beta_2 a$. A univariate comparison obtained from the 2×2 table cross-classifying outcome and group would yield a log odds ratio approximately equal to $\beta_1 + \beta_2(\bar{a}_2 - \bar{a}_1)$. This would incorrectly estimate the effect of group due to the difference in the distribution of age. To account or adjust for this difference, we include age in the model and calculate the logit difference at a common value of age, such as the combined mean, $\bar{a}$. This logit difference is $g(x = 1,\bar{a}) - g(x = 0,\bar{a}) = \beta_1$. Thus, the coefficient β_1 is the log odds ratio which we would expect to obtain from a univariate comparison if the two groups had the same distribution of age.

The data summarized in Table 3.13 provide the basis for an example of interpreting the estimated logistic regression coefficient for a dichotomous variable when the coefficient is adjusted for a continuous variable.

Table 3.13 Descriptive Statistics for Two Groups of 50 Men on AGE and Whether or Not They Had Seen a Physician (PHY) (1 = yes, 0 = no) Within the Last Six Months.

	Group 1		Group 2	
Variable	Mean	SD	Mean	SD
PHY	0.30	0.46	0.80	0.40
AGE	40.18	5.34	48.45	5.02

It follows from the descriptive statistics in Table 3.13 that the univariate log odds ratio for group 2 versus group 1 is $\ln(\hat{\psi}) = \ln(0.80/0.20) - \ln(0.30/0.70) = 2.234$, and the unadjusted estimated odds ratio is $\hat{\psi} = 9.33$. We can also see that there is a considerable difference in the age distribution of the two groups, with men in group 2 being on average nearly 8 years older than those in group 1. We would guess that much of the apparent difference in the proportion of men seeing a physician may be due to age. Analyzing the data with a bivariate model using a coding of 0 for group 1, and 1 for group 2, yields the estimated logistic regression coefficients shown in Table 3.14. The age adjusted log odds ratio is given by the estimated coefficient for group in Table 3.14 and is $\hat{\beta}_1 = 1.559$. The age adjusted odds ratio is $\hat{\psi} = e^{1.559} = 4.75$. Thus, much of the apparent difference between the two groups was in fact due to differences in age.

Let us examine this adjustment in more detail using Figure 3.1. An approximation to the unadjusted odds ratio is obtained by exponentiating the difference $w_2 - w_1$. In terms of the fitted logistic regression model shown in Table 3.14 this difference is

$$[-4.739 + 1.559 + 0.096(48.45)] - [-4.739 + 0.096(40.18)] =$$
$$1.559 + 0.096(48.45 - 40.18)$$

Table 3.14 Results of Fitting the Logistic Regression Model to the Data
Summarized in Table 3.13.

Variable	Estimated Coefficient	Standard Error	Coeff./SE
Group	1.559	0.557	2.80
AGE	0.096	0.048	2.00
Constant	−4.739	1.998	−2.37

Log-likelihood = −53.47

The value of this odds ratio is

$$e^{[1.559 + 0.0956(48.45 - 40.18)]} = 10.48$$

The discrepancy between 10.48 and the actual unadjusted odds ratio, 9.33, is due
to the fact that the above comparison is based on the difference in the average
logit, while the crude odds ratio is approximately equal to a calculation based on
the average estimated logistic probability for the two groups. The age adjusted
odds ratio is obtained by exponentiating the difference $w_4 - w_3$, which is equal
to the estimated coefficient for group. In the example the difference is

$$[-4.739 + 1.559 + 0.096(44.32)] - [-4.739 + 0.096(44.32)] = 1.559$$

The method of adjustment when the variables are all dichotomous,
polytomous, continuous, or a mixture of these is identical to that just described
for the dichotomous–continuous variable case. For example, suppose that
instead of treating age as continuous it was dichotomized using a cutpoint of 45
years. To obtain the age adjusted effect of group we would fit the bivariate
model containing the two dichotomous variables and calculate a logit difference
at the two levels of group and a common value of the dichotomous variable for
age. The procedure would be similar for any number and mix of variables.
Adjusted odds ratios are obtained by comparing individuals who differ only in the
characteristic of interest and have the values of all other variables constant. The
adjustment is statistical as it only estimates what might be expected to be
observed had the subjects indeed differed only on the particular characteristic

being examined, with all other variables having identical distributions within the two levels of outcome.

One point should be kept clearly in mind when interpreting statistically adjusted log odds ratios and odds ratios. The effectiveness of the adjustment is entirely dependent on the adequacy of the assumptions of the model – linearity and constant slope. Departures from these may render the adjustment useless. One such departure, where the relationship is linear but the slopes differ, is called **interaction**. Modeling interactions will be discussed in Section 3.6 and Chapter 4.

3.6 Interaction and Confounding

In the last section we saw how the inclusion of additional variables in a model provides a way of statistically adjusting for potential differences in their distributions. The term confounder is used by epidemiologists to describe a covariate that is associated with both the outcome variable of interest and a primary independent variable or risk factor. When both associations are present then the relationship between the risk factor and the outcome variable is said to be **confounded**. The procedure for adjusting for confounding, described in Section 3.5, is appropriate when there is no interaction. In this section we introduce the concept of interaction and show how we can control for its effect in the logistic regression model. In addition, we will illustrate with an example how confounding and interaction may affect the estimated coefficients in the model.

Interaction can take many different forms, so we begin by describing the situation when it is absent. Consider a model containing a dichotomous risk factor variable and a continuous covariate, such as in the example discussed in Section 3.5. If the association between the covariate (i.e., age) and the outcome variable is the same within each level of the risk factor (i.e., group), then there is no interaction between the covariate and the risk factor. Graphically, the absence of interaction yields a model with two parallel lines, one for each level of the risk factor variable. In general, the absence of interaction is characterized by a model that contains no second or higher order terms involving two or more variables.

When interaction is present, the association between the risk factor and the outcome variable differs, or depends in some way on the level of the covariate. That is, the covariate modifies the effect of the risk factor. Epidemiologists use the term effect modifier to describe a variable that interacts with a risk factor. In the previous example, if the logit is linear in age for the men in group 1, then interaction implies that the logit does not follow a line with the same slope for the second group. In theory, the association in group 2 could be described by almost any model except one with the same slope as the logit for group 1.

The simplest and most commonly used model for including interaction is one in which the logit is also linear in the confounder for the second group, but with a different slope. Alternative models can be formulated which would allow for other than a linear relationship between the logit and the variables in the model within each group. In any model, interaction is incorporated by the inclusion of appropriate higher order terms.

An important step in the process of modeling a set of data is determining whether or not there is evidence of interaction in the data. This aspect of modeling is discussed in Chapter 4. In this section we will assume that when interaction is present it can be modeled by nonparallel straight lines.

Figure 3.2 presents the graphs of three different logits. In this graph, 4 has been added to each of the logits to make plotting more convenient. These logits will be used to further explain what is meant by interaction. Consider an example where the outcome variable is the presence or absence of CHD, the risk factor is gender, and the covariate is age. Suppose that the line labeled l_1 corresponds to the logit for females as a function of age. Line l_2 represents the logit for males. These two lines are parallel to each other, indicating that the effect of age is the same for males and females. In this situation there is no interaction and the log-odds ratios for males versus females, controlling for age, is given by the difference between line l_2 and l_1, $l_2 - l_1$. This difference is equal to the vertical distance between the two lines which is the same for all ages.

Suppose instead that the logit for males is given by the line l_3. Then, for males, the slope of line l_3 is steeper than is the slope of the line l_1 for females, indicating that there is an interaction between age and sex. The estimate of the log-odds ratios for males versus females controlling for age is still given by the vertical distance between the lines, $l_3 - l_1$, but this difference now depends on the age at which the comparison is made. Thus, we cannot estimate the odds ratio for sex without specifying the age at which the comparison is being made. Hence, age is an effect modifier.

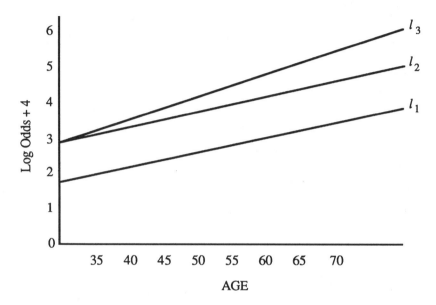

Figure 3.2 Plot of the Logits Under Three Different Models Showing the Presence and Absence of Interaction.

Tables 3.15 and 3.16 present the results of fitting a series of logistic regression models to two different sets of hypothetical data. The variables in each of the data sets are the same: SEX, AGE, and CHD. In addition to the estimated coefficients, the log-likelihood for each model and minus twice the change (deviance) is given. Recall that minus twice the change in the log-likelihood may be used to test for the significance of coefficients for variables added to the model. An interaction is added to the model by creating a variable which is equal to the product of the value of the sex and the value of age. Some programs will automatically create interaction variables, while others require the user to create them through a data modification step.

Table 3.15 Estimated Logistic Regression Coefficients, Log-Likelihood, and the Likelihood Ratio Test Statistic (G) for an Example Showing Evidence of Confounding but No Interaction.

Model	Constant	SEX	AGE	SEX×AGE	Log–Likelihood	G
1	−1.046	1.535			−61.86	
2	−7.142	0.979	0.167		−49.59	24.54
3	−6.103	0.481	0.139	0.059	−49.33	0.52

Examining the results in Table 3.15 we see that the estimated coefficient for the variable SEX changed from 1.535 in model 1 to 0.979 when AGE was added in model 2. Hence, there is clear evidence of a confounding effect due to age. When the interaction term "SEX × AGE" is added in model 3 we see that the change in the deviance is only 0.52 which, when compared to the chi-square distribution with 1 degree of freedom, yields a p-value of 0.47, which is clearly not significant. Note that the coefficient for sex changed from 0.979 to 0.481. This is not surprising since the inclusion of an interaction term, especially when it involves a continuous variable, will usually produce fairly marked changes in the estimated coefficients of dichotomous variables involved in the interaction. Thus, when an interaction term is present in the model we cannot assess confounding via the change in a coefficient. For these data we would prefer to use model 2 which suggests that age is a confounder but not an effect modifier.

Table 3.16 Estimated Logistic Regression Coefficients, Log-Likelihood, and the Likelihood Ratio Test Statistic (G) for an Example Showing Evidence of Confounding and Interaction.

Model	Constant	SEX	AGE	SEX×AGE	Log-Likelihood	G
1	−0.847	2.505			−52.52	
2	−6.194	1.734	0.147		−46.79	11.46
3	−3.105	0.047	0.629	0.206	−44.76	4.06

The results in Table 3.16 show evidence of both confounding and interaction due to age. Comparing model 1 to model 2 we see that the coefficient for sex changes from 2.505 to 1.734. When the age by sex interaction is added to the model we see that the deviance is 4.06, which yields a p-value of 0.04. Since the deviance is significant, we prefer model 3 over model

2, and should regard age as both a confounder and an effect modifier. The net result is that any estimate of the odds ratio for sex should be made with respect to a specific age.

Hence, we see that determining if a covariate, X, is an effect modifier and/or a confounder involves several issues. The plots of the logits shown in Figure 3.2 show us that determining effect modification status involves the parametric structure of the logit, while determination of confounder status involves two things. First the covariate must be associated with the outcome variable. This implies that the logit must have a nonzero slope in the covariate. Second the covariate must be associated with the risk factor. In our example this might be characterized by having a difference in the mean age for males and females. However, the association may be more complex than a simple difference in means. The essence is that we have incomparability in our risk factor groups. This incomparability must be accounted for in the model if we are to obtain a correct, unconfounded, estimate of effect for the risk factor.

In practice, the confounder status of a covariate is ascertained by comparing the estimated coefficient for the risk factor variable from models containing and not containing the covariate. Any "biologically important" change in the estimated coefficient for the risk factor would dictate that the covariate is a confounder and should be included in the model, regardless of the statistical significance of the estimated coefficient for the covariate. On the other hand, we believe that a covariate is an effect modifier only when the interaction term added to the model is both biologically meaningful and statistically significant. When a covariate is an effect modifier, its status as a confounder is of secondary importance since the estimate of the effect of the risk factor depends on the specific value of the covariate.

The concepts of adjustment, confounding, interaction, and effect modification, may be extended to cover the situations involving any number of variables on any measurement scale(s). The dichotomous–continuous variables example illustrated in this section has the advantage that the results may be easily shown graphically. This will not be the case with more complicated situations. The principles for identification and inclusion of confounder and interaction variables into the model are the same regardless of the number of variables and their measurement scales.

3.7 Estimation of Odds Ratios in the Presence of Interaction

In Section 3.6 we showed that when there was interaction between a risk factor and another variable, an estimate of the odds ratio for the risk factor depends on the value of the variable that is interacting with it. In this situation a modification of the formula for estimating odds ratios as given in Section 3.2 is presented which takes into account the correlation between the two interacting variables. For simplicity we will develop the method for a model containing only two variables and their interaction. Extension to a general multivariate model will be discussed and illustrated in Chapter 4.

Consider a model containing a risk factor, F, a covariate, X, and their interaction, $F \times X$. The logit for this model at $F = f$ and $X = x$ is

$$g(f,x) = \beta_0 + \beta_1 f + \beta_2 x + \beta_3 fx \qquad (3.9)$$

The log-odds for $F = f_1$ versus $F = f_0$ with X held constant at $X = x$ is

$$\ln[\psi(F = f_1, F = f_0, X = x)] = g(f_1,x) - g(f_0,x)$$
$$= \beta_1(f_1 - f_0) + \beta_3 x(f_1 - f_0) \qquad (3.10)$$

The estimated log odds is obtained by replacing the parameters in equation (3.10) with their estimated values.

The estimate of the variance of the expression in (3.10) is

$$\hat{\text{var}}\{\ln[\hat{\psi}(F = f_1, F = f_0, X = x)]\} = \hat{\text{var}}(\hat{\beta}_1)(f_1 - f_0)^2$$
$$+ \hat{\text{var}}(\hat{\beta}_3)[x(f_1 - f_0)]^2 + 2\hat{\text{cov}}(\hat{\beta}_1,\hat{\beta}_3)x(f_1 - f_0)^2 \qquad (3.11)$$

Most logistic regression computer packages will estimate the variances and covariances of the estimated parameters in the model. Once the estimates have been obtained they may be substituted into equation (3.11) to obtain an estimate of the variance of the estimated odds ratio. The endpoints of a $100 \times (1 - \alpha)\%$ CI for $\psi(F = f_1, F = f_0, X = x)$ are

$$\exp([\hat{\beta}_1(f_1 - f_0) + \hat{\beta}_3 x(f_1 - f_0)]$$

$$\pm z_{1-\alpha/2}\hat{SE}\{\ln[\hat{\psi}(F = f_1, F = f_0, X = x)]\}) \tag{3.12}$$

These expressions simplify in the case when F is a dichotomous risk factor. If we let $f_1 = 1$ and $f_0 = 0$ then

$$\ln[\hat{\psi}(F = 1, F = 0, X = x)] = \hat{\beta}_1 + \hat{\beta}_3 x \tag{3.13}$$

the estimated variance is

$$\hat{var}\{\ln[\hat{\psi}(F = 1, F = 0, X = x)]\} = \hat{var}(\hat{\beta}_1) + \hat{var}(\hat{\beta}_3)x^2 + 2\hat{cov}(\hat{\beta}_1,\hat{\beta}_3)x \tag{3.14}$$

and the endpoints of the confidence interval for the odds ratio are

$$\exp((\hat{\beta}_1 + \hat{\beta}_3 x) \pm z_{1-\alpha/2}\hat{SE}\{\ln[\hat{\psi}(F = 1, F = 0, X = x)]\}) \tag{3.15}$$

As an example we consider a logistic regression model using the low birth weight data containing the variables AGE and a dichotomous variable, LWD, based on the weight of the mother at the last menstrual period. This variable takes on the value 1 if LWT < 110 pounds, and is zero otherwise. The results of fitting a series of logistic regression models are given in Table 3.17.

Using the estimated coefficient for LWD in model 1 we estimate the odds ratio as $\exp(1.054) = 2.87$. The results shown in the table indicate that AGE is not a strong confounder but does interact with the risk factor, LWD. Thus, to correctly assess the risk of low weight at the last menstrual period, we must include the interaction of this variable with the women's age because the odds ratio is not constant over age. Using equation (3.13) and the results for model 3 the estimated log-odds ratio for low weight at the last menstrual period for a women of AGE = a is

$$\ln[\hat{\psi}(LWD = 1, LWD = 0, AGE = a)] = -1.944 + 0.132a \tag{3.16}$$

Table 3.17 Estimated Logistic Regression Coefficients, Log-Likelihood, the Likelihood Ratio Test Statistic (*G*), and the *p*-Value for the Change for Models Containing LWD and AGE.

Model	Constant	LWD	AGE	LWD×AGE	Log-Likelihood	G	p
0	−0.790				−117.34		
1	−1.054	1.054			−113.12	8.44	0.004
2	−0.027	1.010	−0.044		−112.14	1.96	0.160
3	0.774	−1.944	−0.080	0.132	−110.57	3.14	0.080

In order to obtain the estimated variance we must first obtain the estimated covariance matrix for the estimated parameters. Since this matrix is symmetric most logistic regression software packages will print the results in the form similar to that shown in Table 3.18.

Table 3.18 Estimated Covariance Matrix for the Estimated Parameters in Model 3 of Table 3.17.

Constant	0.828			
LWD	−0.828	2.975		
AGE	−0.353-02	0.353-01	0.157-02	
LWD×AGE	0.353-01	−0.128	−0.157-02	0.573-02
	Constant	LWD	AGE	LWD×AGE

The estimated variance of the log-odds ratio given equation (3.16) is obtained from equation (3.14) and is

$$\hat{\text{var}}\{\ln[\hat{\psi}(\text{LWD} = 1, \text{LWD} = 0, \text{AGE} = a)]\}$$

$$= 2.975 + 0.00573 \times a^2 + 2 \times (-0.128) \times a \qquad (3.17)$$

Values of the estimated odds ratio and 95% CI computed using equations (3.16) and (3.17) for several ages are given in Table 3.19.

Table 3.19 demonstrates that since the log-odds ratio increases linearly in age [equation (3.17)], the effect of LWD on the odds of having a low birth weight baby increases exponentially in age. However, the growing width of the confidence intervals indicates that there is considerable uncertainty in these estimates for women over 30.

Table 3.19 Estimated Odds Ratios and 95% CI for LWT, Controlling for AGE.

AGE	15	20	25	30	35	40	45
$\hat{\psi}$	1.0	2.0	3.9	7.5	14.5	28.1	54.4
$\hat{\psi}_L$	0.2	0.9	1.7	1.9	1.9	1.8	1.7
$\hat{\psi}_U$	3.8	4.2	8.8	29	109	429	1709

3.8 A Comparison of Logistic Regression and a Stratified Analysis for 2 × 2 Tables

Many users of logistic regression, especially those coming from a background in epidemiology, will have performed stratified analyses of 2 × 2 tables to assess interaction and to control confounding. The essential objective of such analyses is to determine whether or not the odds ratios are constant, or homogeneous, over the strata. If they are, a stratified odds ratio estimator such as the Mantel–Haenszel estimator or the weighted logit based estimator will be computed. This analysis may also be performed using the logistic regression modeling techniques discussed in Sections 3.6 and 3.7. In this section we compare these two approaches. An example from the low birth weight data will illustrate the similarities and differences in the two approaches.

Consider an analysis of the risk factor smoking on low birth weight. The crude (or unadjusted) odds ratio computed from the 2 × 2 table shown in Table 3.20, cross-classifying the outcome variable LOW with SMOKE, is $\hat{\psi} = 2.02$.

Table 3.21 presents these data stratifying by the race of the mother. From these tables either the Mantel–Haenszel estimate or the logit based estimate of the odds ratio may be computed.

Table 3.20   Cross-Classification of Low Birth Weight by Smoking Status.

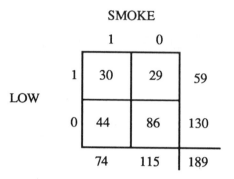

		SMOKE		
		1	0	
LOW	1	30	29	59
	0	44	86	130
		74	115	189

The Mantel–Haenszel estimator is obtained as a weighted average of the stratum specific odds ratios, $\hat{\psi}_i = (a_i \times d_i)/(b_i \times c_i)$, where a_i, b_i, c_i, and d_i are the observed cell frequencies in the 2×2 table for stratum i. For example, $a_1 = 19$, $b_1 = 4$, $c_1 = 33$, and $d_1 = 40$. Denote the total number of subjects in the ith stratum as N_i, for example, $N_1 = 96$. The Mantel–Haenszel estimator is defined in this case by the following formula:

$$\hat{\psi}_{MH} = \frac{\displaystyle\sum a_i \times d_i /N_i}{\displaystyle\sum b_i \times c_i /N_i} \tag{3.18}$$

Evaluating this expression using the data in Table 3.19 yields $\hat{\psi}_{MH} = 13.067/4.234 = 3.09$.

Table 3.21   Cross-Classification of Low Birth Weight by Smoking Status Stratified by RACE.

			RACE							
		White			Black			Other		
SMOKE		1	0	Total	1	0	Total	1	0	Total
LOW	1	19	4	23	6	5	11	5	20	25
	0	33	40	73	4	11	15	7	35	42
Total		52	44	96	10	16	26	12	55	67

Table 3.22 presents the estimated odds ratio, ln(estimated odds ratio), estimate of the variance of the ln(estimated odds ratio) and a weight, w, which is the inverse of the estimate of the variance of ln(estimated odds ratio) for each stratum.

Table 3.22 Tabulation of the Odds Ratios, ln(Estimated Odds Ratios), Estimated Variance of the ln(Estimated Odds Ratios), and the Inverse of the Estimated Variance, w, for Smoking Status Within Each Stratum of RACE.

	White	Black	Other
$\hat{\psi}$	5.758	3.300	1.250
$\ln(\hat{\psi})$	1.751	1.194	0.223
$\mathrm{vâr}[\ln(\hat{\psi})]$	0.358	0.708	0.421
w	2.794	1.413	2.375

The logit based estimator is obtained from a weighted average of the stratum specific log-odds ratio where the weights are the inverse of the variance of the log-odds ratio. The following formula defines this estimator for the current situation,

$$\hat{\psi}_L = \exp[\Sigma w_i \ln(\hat{\psi}_i) \big/ \Sigma w_i] \tag{3.19}$$

Evaluating the logit based estimator using the data in Table 3.22 yields $\hat{\psi}_L = \exp(7.109/6.582) = 2.95$, which is slightly smaller than the Mantel–Haenszel estimate. The high fluctuation in the odds ratio across races suggests that there may be either confounding or effect modification due to RACE, or both. In general, the Mantel–Haenszel estimator and the logit based estimator will be similar when the data are not too sparse within the strata. One considerable advantage of the Mantel–Haenszel estimator is that it may be computed when some of the cell entries are zero.

It is important to note that these estimators provide a correct estimate of the effect of the risk factor only when the odds ratio is constant across the strata. Thus, a crucial step in the analysis is to assess the validity of this assumption. Statistical tests of this assumption are based on a comparison of the stratum specific estimates to an overall estimate computed under the assumption that the

odds ratio is, in fact, constant. The simplest and most easily computed test of
the homogeneity of the odds ratios across strata is based on a weighted sum of
the squared deviations of the stratum specific log-odds ratios from their weighted
mean. This test statistic, in terms of the current notation, is

$$X_H^2 = \sum \{ w_i [\ln(\hat{\psi}_i) - \ln(\hat{\psi}_L)]^2 \} \tag{3.20}$$

Under the hypothesis that the odds ratios are constant, X_H^2 will have a chi-square
distribution with degrees of freedom equal to the number of strata minus 1.
Thus, we would reject the homogeneity assumption when X_H^2 is large.

Using the data in Table 3.22 we have $X_H^2 = 3.017$ which, with 2 degrees of
freedom, yields a p-value of 0.221. This logit based test of homogeneity of
variances suggests that, in spite of the apparent differences in the odds ratios seen
in Table 3.22, the logit based test of homogeneity indicates that they are within
sampling variation of each other. It should be noted that the p-value calculated
from the chi-square distribution will be accurate only when the sample sizes are
not too small within each stratum. This condition holds in this example.

Another test that also may be calculated by hand, but not as easily, is
discussed in Breslow and Day (1980) and has been implemented in SAS-PC.
This test compares the value of a_i to an estimated expected value, $\hat{e}_i$, if the odds
ratio is constant. The formula for the test statistic is

$$X_{BD}^2 = \sum (a_i - \hat{e}_i)^2 / \hat{v}_i \tag{3.21}$$

The quantity $\hat{e}_i$ is obtained as one of the solutions to a quadratic equation given
by the following formula

$$\hat{e}_i = (\hat{\psi}(n_{1i} + m_{1i}) + (n_{0i} - m_{1i}) \pm \{ [\hat{\psi}(n_{1i} + m_{1i})$$
$$+ (n_{0i} - m_{1i})]^2 - [4(\hat{\psi} - 1)(\hat{\psi})(n_{1i} m_{1i})] \}^{1/2}) / 2(\hat{\psi} - 1) \tag{3.22}$$

where $n_{1i} = a_i + b_i$, $m_{1i} = a_i + c_i$ and $n_{0i} = c_i + d_i$. The quantity $\hat{\psi}$ in equation
(3.22) is an estimate of the common odds ratio and either $\hat{\psi}_L$ or $\hat{\psi}_{MH}$ may be
used. The quantity $\hat{v}_i$ is an estimate of the variance of a_i computed under the
assumption of a common odds ratio and is

$$\hat{v}_i = \left(\frac{1}{\hat{e}_i} + \frac{1}{n_{1i} - \hat{e}_i} + \frac{1}{m_{1i} - \hat{e}_i} + \frac{1}{n_{0i} - m_{1i} + \hat{e}_i} \right)^{-1} \qquad (3.23)$$

Using the Mantel–Haenszel estimate, $\hat{\psi}_{MH} = 3.086$ in equations (3.22) and (3.23) the resulting values of $\hat{e}$ and $\hat{v}$ are: $\hat{e}_1 = 17.01$, $\hat{v}_1 = 3.56$, $\hat{e}_2 = 5.91$, $\hat{v}_2 = 1.43$, $\hat{e}_3 = 7.16$, and $\hat{v}_3 = 2.33$. The value of the Breslow–Day statistic obtained is $X^2_{BD} = 3.11$, which is similar to the value of the logit based test.

The same analysis may be performed, and much more easily by fitting three logistic regression models. In model 1 we include only the variable SMOKE. We then add the two design variables for RACE to obtain model 2. For model 3 we add the two RACE × SMOKE interaction terms. The results of fitting these models are given in Table 3.23. Since we are primarily interested in the estimates of the coefficient for SMOKE, the estimates of the coefficients for RACE and the RACE × SMOKE interactions are not shown in Table 3.23.

Table 3.23 Estimated Logistic Regression Coefficients for the Variable SMOKE, Log-Likelihood, the Likelihood Ratio Test Statistic (G), and Resulting p-Value for Estimation of the Stratified Odds Ratio and Assessment of Homogeneity of Odds Ratios Across Strata Defined by RACE.

Model	SMOKE	Log-Likelihood	G	df	p
1	0.704	−114.90			
2	1.116	−109.99	9.83	2	0.007
3	1.751	−108.41	3.16	2	0.206

Using the estimated coefficients in Table 3.23 we have the following estimated odds ratios. The crude odds ratio is $\hat{\psi} = \exp(0.704) = 2.02$. Adjusting for RACE the stratified estimate is $\hat{\psi} = \exp(1.116) = 3.05$. This value is the maximum likelihood estimate of the estimated odds ratio, and it is similar in value to the Mantel-Haenszel estimate ($\hat{\psi}_{MH} = 3.086$) and the logit based estimate ($\hat{\psi}_L = 2.95$). The change in the estimate of the odds ratio from the crude to the adjusted is 2.02 to 3.05, indicating considerable confounding due to RACE.

Assessment of the homogeneity of the odds ratios across the strata is based on the likelihood ratio test of model 2 versus model 3. The value of this statistic from Table 3.23 is $G = 3.156$. This statistic is compared to a chi-square distribution with 2 degrees of freedom, since two interaction terms were added to model 2 to obtain model 3. This value is comparable to the ones from the logit based test, X^2_H, and the Breslow–Day test, X^2_{BD}. Had we used the maximum likelihood estimate of the stratified odds ratio, $\exp(1.116)$, in computing the Breslow–Day test, the resulting statistic would have been equal to the Pearson chi-square goodness-of-fit test of model 2, since model 3 is the saturated model.

The previously described analysis based on likelihood ratio tests may be used when the data have either been grouped into contingency tables such as those shown in Table 3.21 in advance of the analysis, or have remained in casewise form. When the data have been grouped it is possible to point out other similarities between classical analysis of stratified 2×2 tables and an analysis using logistic regression. Day and Byar (1979) have shown that the 1 degree of freedom Mantel–Haenszel test of the hypothesis that the stratum specific odds ratios are 1 is identical to the Score test for the exposure variable when added to a logistic regression model already containing the stratification variable. This test statistic may be easily obtained from a logistic regression package with the capability to perform Score tests such as the EGRET, GLIM, or SAS packages.

Thus, use of the logistic regression model provides a fast and effective way to obtain a stratified odds ratio estimator and to easily assess the assumption of homogeneity of odds ratios across strata.

Exercises

1. Consider the ICU data described in Section 1.5 and use as the outcome variable vital status (STA) and CPR prior to ICU admission (CPR) as a covariate.

    1.1 Demonstrate that the value of the log-odds ratio obtained from the cross-classification of STA by CPR is identical to estimated slope coefficient from the logistic regression of STA on CPR. Verify that the estimated standard error of the estimated slope coefficient for CPR obtained from the logistic regression package is identical to the square root of the sum of the inverse of the cell frequencies from the cross-classification of STA by CPR. Use either set of computations to obtain 95% CI for the odds ratio. What aspect concerning the coding of the variable CPR makes the calculations for the two methods equivalent?

1.2   For purposes of illustration, use a data transformation statement to recode, for this problem only, the variable CPR as follows: 4 = no and 2 = yes. Perform the logistic regression of STA on CPR (recoded). Demonstrate how the calculation of the logit difference of CPR = yes versus CPR = no is equivalent to the value of the log-odds ratio obtained in problem 1.1. Use the results from the logistic regression to obtain the 95% CI for the odds ratio and verify that they are the same limits as obtained in problem 1.1.

2.   Consider the ICU data and use as the outcome variable vital status (STA) and race (RACE) as a covariate.

2.1   Prepare a table showing the coding of the two design variables for RACE using the value RACE = 1 , white, as the reference group. Show that the estimated log-odds ratios obtained from the cross-classification of STA by RACE, using RACE = 1 as the reference group, are identical to estimated slope coefficients for the two design variables from the logistic regression of STA on RACE. Verify that the estimated standard errors of the estimated slope coefficients for the two design variables for RACE are identical to the square root of the sum of the inverse of the cell frequencies from the cross-classification of STA by RACE used to calculate the odds ratio. Use either set of computations to compute the 95% CI for the odds ratios.

2.2   Create design variables for RACE using the method typically employed in ANOVA. Perform the logistic regression of STA on RACE. Show by calculation that the estimated logit differences of RACE = 2 versus RACE = 1 and RACE = 3 versus RACE = 1 are equivalent to the values of the log-odds ratio obtained in problem 2.1. Use the results of the logistic regression to obtain the 95% CI for the odds ratios and verify that they are the same limits as obtained in problem 2.1. Note that the estimated covariance matrix for the estimated coefficients is needed to obtain the estimated variances of the logit differences.

3. Consider the variable AGE in the ICU data set. Prepare a table showing the coding of three design variables based on the empirical quartiles of AGE which uses the first quartile as the reference group. Fit the logistic regression of STA on AGE as recoded into these design variables and plot the three estimated slope coefficients versus the midpoint of the respective age quartile. Plot as a fourth point the value zero at the midpoint of the first quartile of age. Prepare another table showing the coding of three design variables based on the empirical quartiles of AGE which uses the coefficients for the orthogonal polynomials presented in Table 3.11. Fit the logistic regression of STA on AGE as recoded into these design variables. Using the Wald statistics for the coefficients, test for the significance of trends in the logit and compare your results with any trends seen in the plot.

4. Consider the logistic regression of STA on CRN and AGE. Consider CRN to be the risk factor and show that AGE is a confounder of the association of CRN with STA. Addition of the interaction of AGE by CRN presents an interesting modeling dilemma. Examine the main effects only and interaction models graphically. Using the graphical results and any significance tests you feel are needed, select the best model (main effects or interaction) and justify your choice. Estimate relevant odds ratios. Repeat this analysis of confounding and interaction for a model which includes CPR as the risk factor and AGE as the potential confounding variable.

5. Consider an analysis for confounding and interaction for the model with STA as the outcome, CAN as the risk factor, and TYP as the potential confounding variable. Perform this analysis using logistic regression modeling and Mantel–Haenszel analysis. Compare the results of the two approaches.

6. Simulation experiments can often be a useful tool for helping our understanding of relatively complicated processes. The need to include adjustment (confounder) variables and interactions (effect modifiers) in a logistic regression model is one situation where some small simulations can help our understanding of the parameters that drive this process. The following exercises may be performed without great difficulty in most software packages. These exercises are designed to help further intuition and are not to be construed as producing definitive statistical conclusions. We will use some of the previously performed analyses on the data from the ICU study as a focal point for the simulations.

6.1 The logistic regression of STA on CRN and AGE yielded approximately the following estimated logits $g = -1.5 + 1.2 \times CRN$ and $g = -3.3 + 1.0 \times CRN + 0.025 \times AGE$. The inclusion of AGE in the model produced a $20\% = 100[(1.2 - 1.0)/1.0]$ change in the estimated coefficient for CRN. If we consider change of at least 10% as our criteria for an important change, then we would conclude that we need to include AGE in the model. In the ICU study the mean age among those with CRN = 0 is 56.5 and the mean age for those with CRN = 1 is 67.2. This explains the need to include AGE in the model. Would we have seen the same results for a sample size of $n = 100$ or 400? What would have happened had the mean ages been 50 and 65, or 45 and 65, or 55 and 55? We can use the following simulation procedure to get some idea of these effects. Let the variables be c (for CRN), a (for AGE), and y (for outcome) and the sample size be n.

a.   Generate n values of c letting $c = 0$ unless a computer generated uniform (0,1) random variable is less than 0.1

b.   Generate n values of a as $a = 55 + 10 \times c + 18 \times z$ and z is a computer generated normal (0,1) random variable. According to this scheme the values of a will follow a normal distribution with mean 55 and standard deviation 18 within the group with $c = 0$ and mean 65 and the same standard deviation within the group with $c = 1$.

c.   Compute $g = -3.3 + 1.0 \times c + 0.025 \times a$ for each of the n computer generated subjects.

d.   Compute $p = \exp(g)/[1 + \exp(g)]$ for each of the n computer generated subjects.

e.   Compute the values of the outcome variable y as follows: $y = 0$ unless a computer generated uniform (0,1) random variable is less than or equal to p in which case $y = 1$.

f.   Use these data to fit two logistic regression models, one containing c and the other containing c and a. Evaluate the variable a as a confounder. Perform the simulation described in steps a–f using $n = 100, 200$, and 400. Repeat the simulation in step b using $a = 50 + 15 \times c + 18 \times z$, $a = 45 + 20 \times c + 18 \times z$ and $a = 55 + 18 \times z$ with the same sample sizes of $n = 100$, 200, and 400.

6.2   When the CRN by AGE interaction was added to the model the fitted logit became $g = -3.3 + 3.5 \times \mathrm{CRN} + 0.02 \times \mathrm{AGE} - 0.04 \times \mathrm{CRN} \times \mathrm{AGE}$ with a p-value for the interaction term of 0.27. To what extent do the factors of sample size and distribution of age effect the analysis? We may study this by modifying steps c and f above as follows:

c.   Calculate $ca = c \times a$ and $g = -3.3 + 3.5 \times c + 0.02 \times a - 0.04 \times ca$.

f.   Fit three logistic regression models containing respectively the variables c, c and a, and c, a, and ca. Perform this simulation using the same sample sizes and distributions of AGE specified in problem 6.1.

6.3 On the basis of the simulations what factors seem to effect the need to control for confounding or include interactions in a model?

7. We may also use a computer simulation to explore the situation where both variables are dichotomous. The model containing CAN and TYP provides the basis for the parameters in this set of problems.

7.1 Here we proceed using steps a–f of problem 6.1 with the following modifications to steps b and c:

   b. $a = c \times u + (1 - c) \times v$ where u is equal to zero unless a computer generated uniform $(0,1)$ random variable is less than 0.2 and v is equal to zero unless a computer generated uniform $(0,1)$ random variables is less than 0.75.

   c. Calculate $g = -3.8 + 1.3 \times c + 2.7 \times a$. Perform the simulation using the same sample sizes used in problem 6. In addition, in step b of this problem, use as the comparison values $(0.1,0.8)$, $(0.4,0.6)$, and $(0.5,0.5)$ in determining the values of u and v, respectively. Are the factors effecting the need to include a in the model the same as those seen in problem 6?

7.2 To examine interaction we modify step c by calculating $g = -3.6 + c + 2.5 \times a + 0.6 \times c \times a$ and proceeding as in problem 6.2. In addition to the various sample sizes and comparison values given in 7.1, also use in step c the logit $g = -3.6 + c + 2.5 \times a + 1.0 \times c \times a$.

CHAPTER 4

Model-Building Strategies and Methods for Logistic Regression

4.1 Introduction

In the previous chapters we focused on estimating, testing, and interpreting the coefficients in a logistic regression model. The examples discussed were characterized by having few independent variables, and there was perceived to be only one possible model. While there may be situations where this is the case, it is more typical that there are many independent variables that could be included in the model. Hence, we need to develop a strategy and associated methods for handling these more complex situations.

The goal of any method is to select those variables that result in a "best" model within the scientific context of the problem. In order to achieve this goal we must have: (1) a basic plan to follow to select the variables for the model and (2) a set of methods for assessing the adequacy of the model both in terms of the individual variables in the model and from the point of view of the overall fit of the model. We suggest a general method that considers both of these areas.

The methods to be discussed in this chapter are not to be used as a substitute for, but rather, as an addition to clear and careful thought. Successful modeling of a complex data set is part science, part statistical methods, and part experience and common sense. It is our goal to provide the reader with a paradigm that, when applied thoughtfully, will yield the best possible model within the constraints of the available data.

4.2 Variable Selection

The criteria for inclusion of a variable in a model may vary from one problem to the next and from one scientific discipline to another. The traditional approach to statistical model building involves seeking the most parsimonious model that still explains the data. The rationale for minimizing the number of variables in the model is that the resultant model is more likely to be numerically stable, and is more easily generalized. The more variables included in a model, the greater the estimated standard errors become, and the more dependent the model becomes on the observed data. Recently there has been a move among quantitative epidemiologists to include all scientifically relevant variables in the model, irrespective of their contribution to the model. The rationale for this approach is to provide as complete control of confounding as possible within the given data set. This is based on the fact that it is possible for individual variables not to exhibit strong confounding, but when taken collectively, considerable confounding can be present in the data [Miettinen (1976)]. The major problem with this approach is that the model may be overfitted and produce numerically unstable estimates. Overfitting is typically characterized by unrealistically large estimated coefficients and/or estimated standard errors. This may be especially troublesome in problems where the number of variables is large relative to the number of subjects and/or when the overall proportion responding ($y = 1$) is close to either 0 or 1.

There are certain steps one can follow to aid in the selection of variables for a logistic regression model. The whole process is quite similar to the one used in model building in linear regression.

(1) The selection process should begin with a careful univariate analysis of each variable. For nominal, ordinal, and continuous variables with few integer values, we suggest this be done with a contingency table of outcome ($y = 0,1$) versus the k levels of the independent variable. The likelihood ratio chi-square test with $k - 1$ degrees of freedom is exactly equal to the value of the likelihood ratio test for the significance of the coefficients for the $k - 1$ design variables in a univariate logistic regression model that contains that single independent variable. Since the Pearson chi-square test is asymptotically equivalent to the likelihood ratio chi-square test, it may also be used. In addition to the overall test, it is a good idea, for those variables exhibiting at least a moderate level of association, to estimate (along with confidence limits) the individual odds ratios using one of the levels as a reference group.

Particular attention should be paid to any contingency table with a zero cell. This will yield a univariate point estimate for one of the odds ratios of either zero or infinity. Inclusion of such a variable into any logistic regression program will cause one of a number of undesirable numerical outcomes to occur. These will be addressed in the last section of this chapter. It is wise to make a note of the occurrence of the zero cell. Strategies for handling the zero cell include: collapsing the categories of the independent variable in some sensible fashion to eliminate the zero cell; eliminating the category completely; or, if the variable is ordinal scaled, modeling the variable as if it were continuous.

For continuous variables the most desirable univariate analysis involves fitting a univariate logistic regression model to obtain the estimated coefficient, the estimated standard error, the likelihood ratio test for the significance of the coefficient, and the univariate Wald statistic. An alternative analysis, which is equivalent at the univariate level, may be based on the two sample t-test. Descriptive statistics available from a two sample t-test analysis generally include group means, standard deviations, the t statistic, and its p-value. The similarity of this approach to the logistic regression analysis follows from the fact that the univariate linear discriminant function estimate of the logistic regression coefficient is

$$\frac{(\bar{x}_1 - \bar{x}_0)}{s_p^2} = \frac{t}{s_p}\sqrt{\frac{1}{n_1} + \frac{1}{n_0}}$$

and that the linear discriminant function and the maximum likelihood estimate of the logistic regression coefficient are usually quite close when the independent variable is approximately normally distributed within each of the outcome groups, $y = 0, 1$ [see Halpern, Blackwelder, and Verter (1971)]. Thus, univariate analysis based on the t-test should be useful in determining if the variable should be included in the model, since the p-value should be of the same order of magnitude as that of the Wald statistic, Score test, or likelihood ratio test from logistic regression.

For continuous covariates we may wish to supplement the evaluation of the univariate logistic fit with some sort of smoothed scatterplot. This plot, when done on the logit scale, will be helpful in not only ascertaining the potential importance of the variable, but also its appropriate scale. One simple and easily computed form of a smoothed scatterplot was illustrated in Figure 1.2 using the data in Table 1.1. Other more complicated methods which have greater precision are available.

Kay and Little (1986) have illustrated the use of a method proposed by Copas (1983). This method requires that we compute a smoothed value for the response variable for each subject which is a weighted average of the values of the outcome variable over all subjects. The weight for each subject is a continuous decreasing function of the distance of the value of the covariate for the subject under consideration from the value of the covariate for all other cases. For example, for covariate x for the ith subject we would compute

$$\bar{y}_i = \frac{\sum_{j=1}^{n} w(x_j) y_j}{\sum_{j=1}^{n} w(x_j)}$$

where

$$w(x_j) = e^{-(x_j - x_i)^2 / c}$$

and $c > 0$ so that the function tails off at the desired rate. We would plot x_i versus $\bar{y}_i$. Application of this method requires special programs or a data modification capability which will allow nested looping (rereading) of the data. The method illustrated in Figure 1.2 is a discrete analog of this method. A similar graphical analysis may be performed using one of the scatterplot smoothing options available in SYSTAT's graphics module.

Hastie and Tibshirani (1986, 1987) introduced the use of a generalized additive model for the analysis of binary data. The application of their model can provide an effective plotting tool in logistic regression. Unfortunately, application of the method requires that one either obtain their program or develop special modules within the data modification portion of a software package. This is not possible in all software packages. Their method may be used with multivariate models and thus may be more useful for scale identification than the strictly univariate methods such as the method shown in Kay and Little (1986).

(2) Upon completion of the univariate analyses we select variables for the multivariate analysis. Any variable whose univariate test has a p-value < 0.25 should be considered as a candidate for the multivariate model along with all variables of known biologic importance. Once the variables have been identified, we begin with a model containing all of the selected variables.

Use of the 0.25 level as a screening criterion for selection of candidate variables is based on the work by Bendel and Afifi (1977) on linear regression and on the work by Mickey and Greenland (1989) on logistic regression. These authors show that use of a more traditional level (such as 0.05) often fails to identify variables known to be important. Use of the larger level has the disadvantage of including, at the model building stage, variables that are of questionable importance. For this reason, it is important to critically review all variables added to a model before a decision is reached regarding the final model.

One problem with any univariate approach is that it ignores the possibility that a collection of variables, each of which is weakly associated with the outcome, can become an important predictor of outcome when taken together. If this is thought to be a possibility, then we should choose a significance level large enough to allow the suspected variables to become candidates for inclusion in the multivariate model. The best subsets selection technique discussed in Section 4.4 is an effective model-building strategy for identification of collections of variables having this type of association with the outcome variable.

As noted above, the issue of variable selection is made more complicated by different analytic philosophies as well as by different statistical methods. One school of thought argues for the inclusion of all scientifically relevant variables into the multivariate model regardless of the results of univariate analyses. In general, the appropriateness of the decision to begin the multivariate model with all possible variables depends on the overall sample size and the number in each outcome group relative to the total number of candidate variables. When the data are adequate to support such an analysis it may be useful to begin the multivariate modeling from this point. However, when the data are inadequate, this approach can produce a numerically unstable multivariate model. When this occurs the Wald statistics should not be used to select variables because of the unstable nature of the results. In this case we should select a subset of variables based on results of the univariate analyses and refine the definition of "scientifically relevant."

Another approach to variable selection is to use a stepwise method in which variables are selected either for inclusion or exclusion from the model in a sequential fashion based solely on statistical criteria. There are two main versions of the stepwise procedure: (a) forward selection with a test for backward elimination and (b) backward elimination followed by a test for forward selection. The algorithms used to define these procedures in logistic regression will be discussed in Section 4.3. The stepwise approach is useful and intuitively appealing in that it builds models in a sequential fashion and it allows for the examination of a collection of models which might not otherwise have been examined.

An alternative selection method that has not been used extensively in logistic regression is "best subsets selection." With this procedure a number of models containing one, two, three, and so on, variables are examined which are considered the "best" according to some specified criteria. Best subsets linear regression software has been available for a number of years. A parallel theory has been worked out for nonnormal errors models [Lawless and Singhal (1978, 1987a, 1987b)]. We show in Section 4.4 how logistic regression may be performed using any best subsets linear regression program.

Stepwise, best subsets, and other mechanical selection procedures have been criticized because they can yield a biologically implausible model [Greenland (1989)] and can select irrelevant, or noise, variables [Flack and Chang (1987), Griffiths and Pope (1987)]. The problem is not the fact that the computer can select such models, but rather that the analyst fails to carefully scrutinize the resulting model, and reports such results as the final, best model. The wide availability and ease with which stepwise methods can be used has undoubtedly reduced some analysts to a role where they are assisting the computer in model selection rather than the more appropriate alternative. It is only when the analyst understands the strengths, and especially the limitations, of the methods that these methods can serve as a useful tool in the model-building process. The analyst, not the computer, is ultimately responsible for the review and evaluation of the model.

(3) Following the fit of the multivariate model, the importance of each variable included in the model should be verified. This should include (a) an examination of the Wald statistic for each variable and (b) a comparison of each estimated coefficient with the coefficient from the univariate model containing only that variable. Variables that do not contribute to the model based on these criteria should be eliminated and a new model fit. The new model should be compared to the old model through the likelihood ratio test. Also, the estimated coefficients for the remaining variables should be compared to those from the full model. In particular we should be concerned about variables whose coefficients have changed markedly in magnitude. This would indicate that one or more of the excluded variables was important in the sense of providing a needed adjustment of the effect of the variable that remained in the model. This process of deleting, refitting, and verifying continues until it appears that all of the important variables are included in the model and those excluded are either biologically or statistically unimportant.

If the univariate analysis yields an extremely large number of possible variables then it may be worthwhile to employ a stepwise or best subsets method. The objective in using these methods is to identify a fairly full model and then proceed as noted above. For example, we might use forward stepwise selection with a criterion for entry of a p-value less than 0.25 or 0.50.

(4) Once we have obtained a model that we feel contains the essential variables, we should look more closely at the variables in the model and consider the need for including interaction terms among the variables. The question of the appropriate categories for discrete variables should have been addressed at the univariate stage. For continuous scaled variables we should check the assumption of linearity in the logit. The graphs for several different relationships between the logit and a continuous independent variable are shown in Figure 4.1. The figure illustrates the case when the logit is (a) linear, (b) quadratic or nonlinear but continuous, (c) some other nonlinear relationship, and (d) binary where there is a cutpoint above and below which the logit is constant.

The methods we propose for assessing the scale of the logit will provide us with a guide for choosing between these possible models.

Assuming linearity in the logit at the variable selection stage is a common practice and is consistent with the goal of ascertaining if a variable should be in the model. In each of the situations described in Figure 4.1 fitting a linear model would yield a significant slope. Once the variable is identified as important, we can obtain the correct parametric relationship or scale in the model refinement stage. The exception to this would be the rare instance where the response was a symmetric, quadratic, or u-shaped function.

For ease of presentation, we motivate and describe an approach to scale selection using the univariate model. The concept is essentially the same for the multivariate model. In univariate linear regression one method to ascertain whether the model was indeed linear in the variable being fit is to plot the fitted line on the scatterplot of the outcome versus the independent variable and look for any obvious systematic deviations from the line. The effectiveness of this approach will depend to a great extent on the degree of scatter in the plot. A modification of this approach is to break the range of the independent variable into groups and, for each group, to plot the average value of the dependent variable versus the group midpoint.

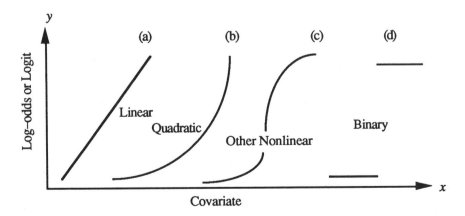

Figure 4.1 Different Types of Models for the Relationship Between the Logit and a Continuous Variable.

To apply this approach in logistic regression we must transform the vertical axis to the logit scale. Thus, in logistic regression we would plot, for each group, the logit of the group mean versus the midpoint of the group. The plot is then examined with respect to the shape of the resulting "curve."

While the concept of this process appears fairly easy and straightforward, its execution involves many computational steps. We need to group the data, compute each group mean, save the group means, transform them to logits, pass these data into a plotting package, and then plot them. Fortunately there is an equivalent procedure which can be easily implemented within any logistic regression package that is based on the following observation. The difference between the logits for two different groups is equal to the value of an estimated coefficient for one of the design variables obtained by fitting a logistic regression model that treats the grouped variable as categorical. Thus, we treat the grouped independent variable as if it were categorical with the lowest group serving as the referent group. Following the fit of the model we plot the estimated coefficients versus the midpoints of the groups and choose the most logical shape for the parametric model. In most applications this "eyeball" assessment will be sufficiently accurate. However, if a more definitive assessment is needed then we can employ the orthogonal design variables discussed in Chapter 3.

An alternative and little known procedure for scale identification in logistic regression is the Box-Tidwell transformation [Box and Tidwell (1962)] for linear regression. The use of this transformation has been examined for use in logistic regression by Guerro and Johnson (1982). The Box-Tidwell approach adds a term of the form $x \ln(x)$ to the model. If the coefficient for this variable is significant we have evidence for non-linearity in the logit. Unfortunately, this procedure has low power in detecting small departures from linearity. However, its simplicity and ease of application may make it worthwhile to try. Kay and Little (1987) illustrate how examination of the marginal distribution of covariates within outcome groups may help suggest the appropriate scale. For example, they show that if the distribution of x is normal with different means and variances, then a linear term, x, and a quadratic term, x^2, will be needed in the logit. If x follows a beta distribution with different parameters within outcome groups, then inclusion of $\ln(x)$ and $\ln(1 - x)$ will be necessary to correctly model x. If x follows a gamma distribution, skewed right, then we need to include x and $\ln(x)$ in the model. The approach of Kay and Little may be most useful for continuous variables when there are enough observations within each outcome group to obtain an accurate approximation to the distribution within each level of the outcome variable.

Once we have ascertained that each of the continuous variables is in the correct scale we begin to check for interactions in the model. In any model an interaction between two variables implies that the effect of one of the variables is not constant over levels of the other. For example, an interaction between sex and age would imply that the slope coefficient for age is different for males and females. The need to include interaction terms in a model is assessed by first creating the appropriate product of the variables in question and then using a likelihood ratio test to assess their significance. By significance we mean interactions must contribute to the model. For example, inclusion of an interaction term in the model whose sole effect is to increase the estimated standard errors without changing the point estimate would not be helpful. In general, for an interaction term to alter both point and interval estimates, the estimated coefficient for the interaction term must attain at least a moderate level of statistical significance. The final decision as to whether an interaction term should be included in a model should be based on statistical as well as practical considerations. That is, the interaction term should also make sense from a biologic perspective.

Example

As an example of the model-building process consider the analysis of the low birth weight data which was used in Chapter 3 to illustrate the interpretation of estimated coefficients in a multivariate model. These data are listed in Appendix 1. In order to make it easier to follow the discussion of model building an abbreviated code sheet is given in Table 4.1 which describes the independent variables and provides the code names for the variables that will be used.

Table 4.1   Code Sheet for the Variables in the Low Birth Weight Data Set. Data Are Listed in Appendix 1.

Variable	Abbreviation
Identification Code	ID
Low Birth Weight (0 = Birth Weight $\geq$ 2500g, 1 = Birth Weight < 2500g)	LOW
Age of the Mother in Years	AGE
Weight in Pounds at the Last Menstrual Period	LWT
Race (1 = White, 2 = Black, 3 = Other)	RACE
Smoking Status During Pregnancy (1 = Yes, 0 = No)	SMOKE
History of Premature Labor (0 = None, 1 = One, etc.)	PTL
History of Hypertension (1 = Yes, 0 = No)	HT
Presence of Uterine Irritability (1 = Yes, 0 = No)	UI
Number of Physician Visits During the First Trimester (0 = None, 1 = One, 2 = Two, etc.)	FTV
Birth Weight in Grams	BWT

Low birth weight is an outcome that has been of concern to physicians for years. This is due to the fact that infant mortality rates and birth defect rates are very high for low birth weight babies. A woman's behavior during pregnancy (including diet, smoking habits, and receiving prenatal care) can greatly alter the chances of carrying the baby to term and, consequently, of delivering a baby of normal birth weight.

The variables identified in the code sheet given in Table 4.1 have been shown to be associated with low birth weight in the obstetrical literature. The goal of the current study was to ascertain if these variables were important in the population being served by the medical center where the data were collected. The results of fitting the univariate logistic regression models to these data are given in Table 4.2. In this table, for purposes of a preliminary analysis only, the variables recording the number of physician visits during the first trimester, and the number of previous premature labors have been treated as if they were continuous variables.

Table 4.2 Univariate Logistic Regression Models for Low Birth Weight Data.

Variable	$\hat{\beta}$	$\hat{S}E(\hat{\beta})$	$\hat{\psi}$	95% CI	Log-Likelihood	G	p
Constant	−0.790	0.157			−117.34		
AGE	−0.051	0.031	0.60	(0.32,1.11)	−115.96	2.76	0.10
LWT	−0.014	0.006	0.87	(0.77,0.98)	−114.35	5.98	0.02
RACE (1)	0.845	0.463	2.33	(0.94,5.77)	−114.83	5.01	0.08
RACE (2)	0.636	0.347	1.89	(0.96,3.73)			
SMOKE	0.704	0.319	2.02	(1.08,3.78)	−114.90	4.87	0.03
PTL	0.802	0.316	2.23	(1.20,4.14)	−113.95	6.78	0.01
HT	1.214	0.607	3.37	(1.02,11.06)	−115.33	4.02	0.04
UI	0.947	0.417	2.58	(1.14,5.83)	−114.80	5.08	0.02
FTV	−0.135	0.156	0.87	(0.64,1.19)	−116.95	0.77	0.38

In Table 4.2 we present, for each variable listed in the first column, the following information: (1) the estimated slope coefficient(s) for the univariate logistic regression model containing only this variable, (2) the estimated standard error of the estimated slope coefficient, (3) the estimated odds ratio, which is obtained by exponentiating the estimated coefficient [Note: For the variables AGE and weight at last menstrual period (LWT) the odds ratio is for an increase of 10 years and 10 pounds, respectively. This was done since a change of 1 year or 1 pound would not be as meaningful.], (4) the 95% CI for the odds ratio, (5) the value of the log-likelihood for the model, and (6) the likelihood ratio test statistic, G, for the hypothesis that the slope coefficient is zero. This is obtained as minus twice the difference between the log-likelihoods for the constant only model given in the first row and the model containing the respective variable. Under the null hypothesis this quantity will follow the chi-square distribution with 1 degree of freedom, except for the variable RACE, where it will have 2 degrees of freedom.

With the exception of the number of first trimester physician visits (FTV), there is evidence that each of the variables has some association with the outcome, low birth weight. We base this observation on an inspection of the confidence interval estimates which, for most variables, either do not contain 1 or just barely do. In addition the values for the likelihood ratio tests each exceed the 90th percentile of their respective chi-square distributions. Thus, based on the univariate results, we begin the multivariate model with all the variables except FTV. A detailed analysis of FTV obtained from the design variables for a grouping of 0, 1, 2, 3, ≥4 confirmed that there was no association between the number of first trimester physician visits and low birthweight in these data. The results of fitting the multivariate model are given in Table 4.3.

Table 4.3 Estimated Coefficients, Estimated Standard Errors, and Coefficient/Standard Error for the Multivariate Model Containing Variables Identified in the Univariate Analysis.

Variable	Estimated Coefficient	Estimated Standard Error	Coeff./SE
AGE	−0.027	0.036	−0.74
LWT	−0.015	0.007	−2.19
RACE (1)	1.263	0.525	2.40
RACE (2)	0.862	0.438	1.97
SMOKE	0.923	0.400	2.31
PTL	0.542	0.345	1.57
HT	1.834	0.690	2.66
UI	0.759	0.459	1.65
Constant	0.464	1.201	0.38

Log-likelihood = −100.71

On the basis of the output displayed in Table 4.3 it appears that all of the variables except for AGE demonstrate considerable importance in the multivariate model. This suggests that we fit a new model which does not contain AGE. The results of fitting this model are given in Table 4.4.

The likelihood ratio test for the difference between the models in Tables 4.3 and 4.4 (a test for the significance of AGE) yields a value of $G = -2[-100.99 - (-100.71)] = 0.56$. Comparing this to a chi-square distribution with 1 degree of freedom yields a p-value of 0.45, demonstrating that the variable AGE adds little to the model once the other variables have been included in the model. This is further supported by the fact that the values of the estimated coefficients for the other variables are nearly identical in the two models.

Table 4.4 Estimated Coefficients, Estimated Standard Errors, and Coefficient/Standard Error for the Multivariate Model Containing Variables Significant in the Model in Table 4.3.

Variable	Estimated Coefficient	Estimated Standard Error	Coeff./SE
LWT	−0.016	0.007	−2.33
RACE (1)	1.326	0.521	2.54
RACE (2)	0.897	0.432	2.07
SMOKE	0.939	0.397	2.36
PTL	0.503	0.340	1.48
HT	1.855	0.693	2.67
UI	0.786	0.456	1.72
Constant	−0.087	0.949	−0.09

Log-likelihood = −100.99

At this point in the model development we have to make a decision concerning the variable AGE. AGE is known to be a biologically important variable, yet its coefficient is not statistically significant in this low birth weight model.

For some adverse reproductive outcomes and for some populations it has been shown that the logit, as a function of age, is quadratic or "U" shaped. As noted earlier, with occurrence of such a shape, a linear fit may have zero slope. To examine if this was the situation in these data we formed three design variables based on the quartiles of AGE, and replaced AGE (continuous) with these design variables in the model. A summary of the results of this fit are presented in Table 4.5 In addition to the estimated logistic regression coefficients, Table 4.5 shows the midpoint of each quartile, the proportion of low birth weights (quartile average of the dependent variable LOW), and point and interval estimates for the odds ratio obtained from the estimated coefficients.

The estimated coefficients and odds ratios in Table 4.5 show that there is no significant linear or quadratic trend. In particular, the fourth quartile does not have an estimated odds ratio greater than that of the third quartile, which would be necessary to support a quadratic or "U" shaped function. The second and fourth quartiles have similar estimates and confidence intervals. Addition of the variable AGE[ln(AGE)] to the model containing AGE as continuous did not yield a significant coefficient for AGE[ln(AGE)]. This supports treating AGE as linear in the logit. Because AGE is known to interact with other variables, we will keep AGE in the model as continuous and linear in the logit until we explore the possibility of interactions. Furthermore, its presence in the model does not adversely affect the estimated standard errors of the other variables, as can be seen by comparing the values in Tables 4.3 and 4.4.

Table 4.5 Results of the Quartile Analyses of AGE from the Multivariate Model Containing the Variable Shown in the Model in Table 4.3.

Quartile	1	2	3	4
Midpoint	16.5	21	24.5	35.5
Number	35	59	41	54
Mean	0.34	0.31	0.39	0.24
Estimated Coefficient	0	−0.151	0.390	−0.108
$\hat{\psi}$	1.0	0.86	1.48	0.89
95% CI	−	(0.32,2.34)	(0.52,4.22)	(0.31,2.60)

Before proceeding to assess interactions we need to examine the variables that have been modeled as continuous to obtain the correct scale in the logit. The variables we need to check are weight at last menstrual period (LWT), and number of previous premature labors (PTL).

We begin by determining the quartiles of the distribution of LWT and create three design variables using the lowest quartile as the reference group. These design variables are then used in the multivariate model in place of LWT. The format of Table 4.6 is identical to Table 4.5 which displays the results of the quartile analysis for AGE.

To ascertain the scale of LWT we examine the estimated coefficients in Table 4.6. If the logit was linear in LWT then we would expect to see either a linear increasing or decreasing trend in the estimated coefficients. What we see in Table 4.6 is not evidence of linearity; instead, a binary model is suggested. The three estimated coefficients are similar in magnitude and the confidence intervals have a great deal of overlap.

Table 4.6 Results of the Quartile Analysis of LWT from the Multivariate Model Containing the Variables Shown in the Model in Table 4.3.

Quartile	1	2	3	4
Midpoint	95	115.5	129.5	195
Number	42	50	47	50
Mean	0.5	0.26	0.28	0.24
Estimated Coefficient		−0.807	−0.694	−1.039
$\hat{\psi}$		0.44	0.49	0.35
95% CI		(0.18,1.13)	(0.19,1.31)	(0.13,1.00)

These observations suggest that we should create a dichotomous variable taking on the value 1 if LWT is in the first quartile and zero otherwise. The value of 1 is assigned to the first quartile because that group is at increased risk relative to the other three quartiles. This can be seen in Table 4.6 where all the estimated coefficients are negative and the estimated odds ratios are less than 1. The results using the new dichotomous variable, LWD, in the multivariate model are given in Table 4.7.

Table 4.7 Results of Using the Dichotomous Variable, LWD, in the Multivariate Model.

Group	1	2
Midpoint	129.5	95
Number	147	42
Mean	0.26	0.5
Estimated Coefficient		0.830
$\hat{\psi}$		2.29
95% CI		(1.05,5.01)

The results in Table 4.7 show that the women in the first quartile (LWT < 110) have a greater risk of having a low birth weight baby than women whose weight at the time of the last menstrual period was greater than 110 pounds. Thus, the results of our examination of the scale of the variable LWT have clarified our understanding of the association between mother's weight at last menstrual period and low birth weight.

The scale of the variable, number of previous premature labors, PTL, is simplified by the fact that of the 189 subjects, 159 had no prior history, 24 had one, and 6 had two or three. In this case we have little choice but to dichotomize the variable into a "yes – no" variable, since the frequency is so low in the last group. Therefore, we create a new variable PTD with a value of zero if PTL is zero and 1 otherwise.

The results of fitting the multivariate model with the new variables LWD and PTD are given in Table 4.8.

Table 4.8 Estimated Coefficients, Estimated Standard Errors, and Coefficients/Standard Errors for the Multivariate Model Containing LWD and PTD, Dichotomous Variables Created from LWT and PTL.

Variable	Estimated Coefficient	Estimated Standard Error	Coeff./SE
AGE	−0.046	0.037	−1.25
LWD	0.842	0.405	2.08
RACE (1)	1.073	0.514	2.09
RACE (2)	0.815	0.444	1.84
SMOKE	0.807	0.404	2.00
PTD	1.282	0.461	2.78
HT	1.435	0.647	2.22
UI	0.658	0.466	1.41
Constant	−1.217	0.954	−1.28

Log-likelihood = −98.78

The multivariate model we began with in Table 4.3 had a log-likelihood of −100.71, and the model in Table 4.8 has a log-likelihood of −98.78, a change of 1.93. We cannot base a test on this value since the models in the two tables are not nested (i.e., the variables in one model are not a proper subset of the variables in the second). We now turn to the question of interactions among the variables in the model.

A total of 42 interactions may be formed from the variables in the model in Table 4.8. In situations such as this we recommend that only those interactions which we have prior reason to be concerned about or which make biologic sense be formed and investigated. In the current model the variables AGE, RACE, and smoking status (SMOKE) are each known to have the potential to interact with many variables. Hence, we will examine the interaction of each of these with all other variables in the model. It is known that weight and hypertension are related so we will also check for an interaction between these two. At this stage in the model-building process we are interested in assessing the contribution (significance) of each interaction to the previously developed multivariate model (Table 4.8). The results of adding each interaction to this model are presented in Table 4.9. We are interested in assessing how inclusion of interaction terms would alter the point and interval estimates computed from the main effects model in Table 4.8. As a general rule an interaction must demonstrate at least a moderate level of statistical significance for this to occur. Hence, for screening purposes it suffices to present the log-likelihood, the likelihood ratio test statistic, G, for the current model versus the main effects only model, the degrees of freedom for this test, and its p-value.

In Table 4.9 we see that only two interactions are worth pursuing further. These are the AGE × LWD interaction and the SMOKE × LWD interaction. An additional model containing both of these interaction terms was fit. The results of fitting this model are given in the last line of Table 4.9. This model seems to provide a significant improvement over the main effects only model. The inclusion of these interaction terms into the model offers the possibility to better describe the effects of age, low weight of the mother at last menstrual period, and smoking on giving birth to a low birth weight baby. The estimated coefficients for this model are given in Table 4.10.

We now turn to the interpretation of the fitted model in Table 4.10. Usually we would assess the overall fit and adequacy of the model before proceeding with inferences based upon the model. Because discussion of methods for assessment of fit is deferred until Chapter 5, we will proceed as if this step were done and the model judged to be adequate.

Inferences from a fitted logistic regression model ordinarily begin with estimation of odds ratios for the various risk factors in the model. For variables not involving interactions these are obtained by exponentiation of estimated coefficients. In the current model this may be applied to the variables RACE, hypertension (HT), uterine irritability (UI), and previous premature delivery

(PTD). A modification of this approach is required to obtain estimated odds ratios for smoking status (SMOKE), low weight of the mother at last menstrual period (LWD), and AGE since each of these variables are included in an interaction. Following the estimation of the odds ratios for the "main effects" only variables, we present in detail the extension of the procedure given in Section 3.7 for estimating odds ratios in the presence of interaction for the current more complicated model.

Table 4.9 Log-likelihood, Likelihood Ratio Test Statistic (G), Degrees of Freedom (df), and p-Value for Possible Interactions of Interest to Be Added to the Main Effects Only Model.

Interaction	Log-Likelihood	G	df	p-value
Main Effects Only*	−98.78			
AGE ×RACE	−98.53	0.50	2	0.78
AGE ×SMOKE	−98.51	0.54	1	0.46
AGE × HT	−98.39	0.78	1	0.38
AGE ×UI	−98.76	0.04	1	0.84
AGE ×LWD	−97.50	2.56	1	0.11
AGE ×PTD	−98.36	0.84	1	0.36
RACE ×SMOKE	−97.61	2.34	2	0.31
RACE ×HT	−98.63	0.30	2	0.86
RACE ×UI	−97.62	2.32	2	0.31
RACE ×LWD	−97.08	3.40	2	0.18
RACE ×PTD	−98.50	0.56	2	0.76
SMOKE × HT	−98.71	0.14	1	0.71
SMOKE × UI	−98.12	1.32	1	0.25
SMOKE ×LWD	−97.61	2.34	1	0.13
SMOKE ×PTD	−98.31	0.94	1	0.33
LWD ×HT	−98.22	1.12	1	0.30
AGE ×LWD + SMOKE ×LWD	−96.01	5.54	2	0.06

*Main effects model from Table 4.8.

Estimated odds ratios and 95% CI for the variables RACE, HT, UI, and PTD are given in Table 4.11. These confidence intervals suggest that each of these variables effects low birth weight. Even though the interval for UI includes 1, it is sufficiently skewed to the right to suggest that the association of

UI on low birth weight, after controlling for the other variables in the model, may be considerable.

Table 4.10 Estimated Coefficients, Estimated Standard Errors and Coefficients/Standard Errors for the Multivariate Model Containing Main Effects and Significant Interactions.

Variable	Estimated Coefficient	Estimated Standard Error	Coeff./SE
AGE	−0.084	0.046	−1.84
RACE (1)	1.083	0.519	2.09
RACE (2)	0.760	0.460	1.63
SMOKE	1.153	0.458	2.52
HT	1.359	0.662	2.05
UI	0.728	0.480	1.52
LWD	−1.730	1.868	−0.93
PTD	1.232	0.471	2.61
AGE ×LWD	0.147	0.083	1.78
SMOKE×LWD	−1.407	0.819	−1.72
Constant	−0.512	1.088	−0.47

Log-likelihood = −96.01

We now turn to an analysis of the variables SMOKE, LWD, and AGE. Several different odds ratios are of interest. We would like to estimate the odds ratio for LWD among both smokers and nonsmokers, controlling for age. We should also estimate the odds ratio for smoking separately among those women of low weight (LWD = 1) and those women not of low weight (LWD = 0). A fifth odds ratio of interest is the odds for both smoking and low weight (SMOKE = 1, LWD = 1) relative to nonsmoking women not of low weight (SMOKE = 0, LWD = 0), controlling for age. Each of these odds ratios is estimated by exponentiation of an appropriately calculated logit difference. We first present the logit and the differences in general terms, and then use the values of the estimated coefficients in Table 4.10 to get numeric values.

Table 4.11 Estimated Odds Ratios and 95% CI for the Variables RACE, HT, UI, and PTD.

Variable	$\hat{\psi}$	95% CI
RACE (White)	1.0	
RACE (Black)	3.0	(1.1,8.2)
RACE (Other)	2.1	(0.9,5.3)
HT	3.9	(1.1,14.2)
UI	2.1	(0.8,5.3)
PTD	3.4	(1.4,8.6)

To simplify the presentation we renumber the coefficients in the model. Let "1" denote the subscript for the coefficient for the variable LWD, let "2" denote the subscript for the coefficient for the variable SMOKE , let "3" denote the subscript for the coefficient for the variable AGE, let "4" denote the subscript for the coefficient for the interaction of SMOKE and LWD, and let "5" denote the subscript for the coefficient for the interaction of LWD and AGE. For example, the estimated coefficient for LWD is $\hat{\beta}_1 = -1.730$ and the estimated coefficient for the interaction SMOKE $\times$ LWD is $\hat{\beta}_4 = -1.407$. The remaining variables in the model will remain constant when we estimate the odds ratios and therefore their contribution to the logit will be subtracted out when the logit difference is computed. We will represent their contribution to the logit by the term $\hat{\beta}'z$. This term will contain the contribution of RACE, HT, UI, and PTD to the logit. Using this notation, the estimated logit for a woman with LWD = w, SMOKE = s, and AGE = a is

$$\hat{g}(\text{LWD} = w, \text{SMOKE} = s, \text{AGE} = a, z)$$

$$= \hat{\beta}_0 + \hat{\beta}_1 w + \hat{\beta}_2 s + \hat{\beta}_3 a + \hat{\beta}_4 w \times s + \hat{\beta}_5 w \times a + \hat{\beta}'z \qquad (4.1)$$

We use this to express the logit for the four different combinations of values of LWD and SMOKE. We may obtain the values of the estimated logit differences by substituting the values of the estimated parameters from Table 4.9 into the expressions in Table 4.12. The estimated odds ratios are obtained by exponentiation of these values. These computations are presented in Table 4.13 for a woman of AGE = 30 years. In Table 4.13 we use the symbol "L" to represent LWD = 1 and "NL" to represent LWD = 0 while the symbols "S" and "NS" are used for the two values of the smoking status variable.

Table 4.12 Expression of the Logits and Logit Differences in Terms of the Estimated Parameters for the Four Possible Combinations of the Values of LWD and SMOKE for a Women of AGE a.

LWD

SMOKE	0	1	Logit Difference
0	$\hat{\beta}_0 + \hat{\beta}_3 a + \boldsymbol{\beta}'\mathbf{z}$	$\hat{\beta}_0 + \hat{\beta}_1 + \hat{\beta}_3 a + \hat{\beta}_5 a + \boldsymbol{\beta}'\mathbf{z}$	$\hat{d}_1 = \hat{\beta}_1 + \hat{\beta}_5 a$
1	$\hat{\beta}_0 + \hat{\beta}_2 + \hat{\beta}_3 a + \boldsymbol{\beta}'\mathbf{z}$	$\hat{\beta}_0 + \hat{\beta}_1 + \hat{\beta}_2 + \hat{\beta}_3 a + \hat{\beta}_4 + \hat{\beta}_5 a + \boldsymbol{\beta}'\mathbf{z}$	$\hat{d}_2 = \hat{\beta}_1 + \hat{\beta}_4 + \hat{\beta}_5 a$
Logit Difference	$\hat{d}_3 = \hat{\beta}_2$	$\hat{d}_4 = \hat{\beta}_2 + \hat{\beta}_4$	$\hat{d}_5 = \hat{\beta}_1 + \hat{\beta}_2 + \hat{\beta}_4 + \hat{\beta}_5 a$

Table 4.13 Values of the Estimated Logit Differences, Odds Ratios, and 95% CI for a Women of AGE 30.

Effect	Among	Logit Difference	$\hat{\psi}$	95% CI
L	NS	$-1.730 + 0.147 \times 30$	14.76	(2.4,89.0)
L	S	$-1.730 - 1.407 + 0.147 \times 30$	3.57	(0.7,18.0)
S	NL	1.153	3.17	(1.3,7.8)
S	L	$1.153 - 1.407$	0.76	(0.2,3.3)
S + L		$-1.730 + 1.153 - 1.407 + 0.147 \times 30$	11.31	(2.1,61.6)

Calculation of the confidence limits requires that we obtain estimates of the variance for each of the linear combinations of the coefficients in the expressions for the logit differences in Table 4.13. These calculations, while somewhat tedious, are not difficult. They follow basic rules for calculating the variance of a linear combination of related variables. These rules specify that the variance of a linear combination is the sum of the variance of each term in the expression plus two times the covariance of each possible pair of terms. The expressions for each of the five logit differences are given below:

$$\text{var}(\hat{d}_1) = \text{var}(\hat{\beta}_1) + a^2\text{var}(\hat{\beta}_5) + 2a\,\text{cov}(\hat{\beta}_1,\hat{\beta}_5) \tag{4.2}$$

$$\text{var}(\hat{d}_2) = \text{var}(\hat{\beta}_1) + \text{var}(\hat{\beta}_4) + a^2\text{var}(\hat{\beta}_5) + 2\text{cov}(\hat{\beta}_1,\hat{\beta}_4)$$
$$+ 2a\,\text{cov}(\hat{\beta}_1,\hat{\beta}_5) + 2a\,\text{cov}(\hat{\beta}_4,\hat{\beta}_5) \tag{4.3}$$

$$\text{var}(\hat{d}_3) = \text{var}(\hat{\beta}_2) \tag{4.4}$$

$$\text{var}(\hat{d}_4) = \text{var}(\hat{\beta}_2) + \text{var}(\hat{\beta}_4) + 2\text{cov}(\hat{\beta}_2,\hat{\beta}_4) \tag{4.5}$$

and

$$\text{var}(\hat{d}_5) = \text{var}(\hat{\beta}_1) + \text{var}(\hat{\beta}_2) + \text{var}(\hat{\beta}_4) + a^2\text{var}(\hat{\beta}_5)$$
$$+ 2\text{cov}(\hat{\beta}_1,\hat{\beta}_2) + 2\text{cov}(\hat{\beta}_1,\hat{\beta}_4) + 2a\,\text{cov}(\hat{\beta}_1,\hat{\beta}_5)$$
$$+ 2\text{cov}(\hat{\beta}_2,\hat{\beta}_4) + 2a\,\text{cov}(\hat{\beta}_2,\hat{\beta}_5) + 2a\,\text{cov}(\hat{\beta}_4,\hat{\beta}_5) \tag{4.6}$$

where var stands for variance and cov stands for covariance. Estimates are calculated by first obtaining the covariance matrix from the software package being used and then substituting the estimated values into equations (4.2)–(4.6). The portion of the full 11 by 11 covariance matrix containing the estimates of the variances and covariances for only the estimated parameters used in the logit differences in Table 4.13 is provided in Table 4.14. The entries in Table 4.14 are presented in the lower triangular matrix form used by many programs. The entry at the top of each column is the estimate of the variance of the estimated coefficient for the term given in the column. The remainder of the values are estimates of covariances. A particular covariance is found by intersecting the row and column containing the variables of interest. For example, the estimated variance of the estimated coefficient for AGE is $\hat{var}(\hat{\beta}_3) = 0.0021$. The estimate of the covariance for the estimated coefficients of LWD and SMOKE is $\hat{cov}(\hat{\beta}_1,\hat{\beta}_2) = 0.0063$.

Table 4.14 Estimates of the Variances and Covariances for the Estimated Parameters Used in the Calculation of the Logit Differences Used to Compute the Odds Ratios for LWD and Smoking Adjusting for AGE.

	LWD	SMOKE	AGE	LWD×SMOKE	LWD×AGE
LWD	3.490				
SMOKE	0.626-02	0.210			
AGE	0.447-01	−0.863-02	0.208-02		
LWD × SMOKE	−0.120	−0.165	0.776-03	0.670	
LWD × AGE	−0.147	0.284-02	−0.203-02	−0.100-01	0.687-02

We now turn to the interpretation of the estimated odds ratios in Table 4.13. We see that low weight (at the last menstrual period) in itself is an important risk factor when the woman does not smoke; the estimated odds ratio is 14.76. Among smokers the odds ratio for low weight is not as impressive, but the highly skewed confidence interval suggests that the effect, while not statistically significant, may be important. This difference in the odds ratio for LWD for smokers as compared to nonsmokers results from the negative interaction between smoking and LWD.

We see that SMOKING is also a significant risk factor among women who are not of low weight with an estimated odds ratio of 3.17. The odds ratio for smoking among women of low weight is 0.76, which with confidence limits of (0.2,3.3) would be judged to be both biologically as well as statistically not significant. Taken together, the odds ratio is estimated to be 11.45 for a smoking, low weight woman, relative to a nonsmoking woman not of low weight.

We see from this analysis that, after controlling for smoking group, low weight of the mother is an important variable in the prediction of low birth weight. This is rather surprising since the Wald test statistics presented in the last column of Table 4.10 suggest that LWD is the least significant of the variables studied. Hence, studying odds ratios within categories of other covariates is an important adjunct to the logistic regression analysis, particularly when there is a sizable interaction term in the model.

4.3 Stepwise Logistic Regression

Stepwise selection of variables has been widely used in linear regression. Most major software packages have either a separate program or an option to perform this type of analysis. At one time, stepwise regression was an extremely popular method for model building. However, in recent years there has been a shift away from deterministic methods for model building to purposeful selection of variables. Methodology for performing stepwise logistic regression has been available for much less time [Hosmer, Wang, Lin, and Lemeshow (1978)]. Among major software packages only BMDP offers a program for stepwise logistic regression. We feel that the procedure provides a useful and effective data analysis tool. In particular, there are times when the outcome being studied is relatively new (e.g., AIDS) and the important covariates may not be known and associations with the outcome not well understood. In these instances most studies will collect many possible covariates and screen them for significant associations. Employing a stepwise selection procedure can provide a fast and effective means to screen a large number of variables, and to simultaneously fit a number of logistic regression equations.

Any stepwise procedure for selection or deletion of variables from a model is based on a statistical algorithm which checks for the "importance" of variables, and either includes or excludes them on the basis of a fixed decision rule. The "importance" of a variable is defined in terms of a measure of the statistical significance of the coefficient for the variable. The statistic used depends on the assumptions of the model. In stepwise linear regression an F-test is used since the errors are assumed to be normally distributed. In logistic regression the errors are assumed to follow a binomial distribution, and significance is assessed via the likelihood ratio chi-square test. Thus, at any step in the procedure the most important variable, in statistical terms, will be the one that produces the greatest change in the log-likelihood relative to a model not containing the variable (i.e., the one that would result in the largest likelihood ratio statistic, G).

We have pointed out that a polytomous variable with k levels is appropriately modeled through its $k - 1$ design variables. Since the magnitude of G depends on its degrees of freedom, any procedure based on the likelihood ratio test statistic, G, must account for possible differences in degrees of freedom between variables. This is done by assessing significance through the p-value for G.

We will describe and illustrate the algorithm for forward selection followed by backward elimination in stepwise logistic regression. Any variants of this algorithm are simple modifications of this procedure. The method will be described by considering the statistical computations that the computer must perform at each step of the procedure.

Step (0): Suppose we have available a total of p possible independent variables, all of which are judged to be of plausible "biologic" importance in studying the outcome variable. Step (0) begins with a fit of the "intercept only model" and an evaluation of its log-likelihood, L_0. This is followed by fitting each of the p possible univariate logistic regression models and comparing their respective log-likelihoods. Let the value of the log-likelihood for the model containing variable x_j at step zero be denoted by $L_j^{(0)}$. The subscript j refers to that variable which has been added to the model, and the superscript (0) refers to the step. This notation will be used throughout the discussion of stepwise logistic regression to keep track of both step number and variables in the model.

Let the value of the likelihood ratio test for the model containing x_j versus the intercept only model be denoted by $G_j^{(0)} = 2(L_j^{(0)} - L_0)$, and its p-value be denoted by $p_j^{(0)}$. Hence, this p-value is determined by the tail probability $\Pr[\chi^2(\nu) > G_j^{(0)}] = p_j^{(0)}$, where $\nu = 1$ if x_j is continuous and $\nu = k - 1$ if x_j is polytomous with k categories.

The most important variable is the one with the smallest p-value. If we denote this variable by x_{e_1}, then $p_{e_1}^{(0)} = \min(p_j^{(0)})$, where "min" stands for selecting the minimum of the quantities enclosed in the brackets. The subscript "e_1" is used to denote that the variable is a candidate for entry at step 1. For example, if variable x_2 had the smallest p-value, then $p_2^{(0)} = \min(p_j^{(0)})$, and

$e_1 = 2$. Just because x_{e_1} is the most important variable, there is no guarantee that it will be "statistically significant." For example, if $p_{e_1}^{(0)} = 0.83$, we would probably conclude that there is little point in continuing this analysis because the "most important" variable is not related to the outcome. On the other hand, if $p_{e_1}^{(0)} = 0.003$, we would like to look at the logistic regression containing this variable and then see if there are other variables which are important given that x_{e_1} is in the model.

A crucial aspect of using stepwise logistic regression is the choice of an "alpha" level to judge the importance of variables. Let p_E denote our choice where the "E" stands for entry. The choice for p_E will determine how many variables will eventually be included in the model. Bendel and Afifi (1977) have studied the choice of p_E for stepwise linear regression, and Costanza and Afifi (1979) have studied the choice for stepwise discriminant analysis. The results of this research have shown that the choice of $p_E = 0.05$ is too stringent, often excluding important variables from the model. Choosing a value for p_E in the range 0.15 to 0.20 is more highly recommended. While previous research considered only normal theory models (i.e., linear regression or discriminant analysis), there is reason to believe that use of p_E in the same range would be a suitable criterion for stepwise logistic regression since logistic regression may be viewed as an offshoot of the normal theory discriminant function model. Moreover, use of p_E in this range will provide some assurance that the stepwise procedure will select variables whose coefficients are different from zero.

Sometimes the goal of the analysis may be broader, and models containing more variables are sought to provide a more complete picture of possible models. In these cases use of $p_E = 0.25$ might be a reasonable choice. Whatever the choice for p_E, a variable will be judged important enough to include in the model if the p-value for G is less than p_E. Thus, the program proceeds to step (1) if $p_{e_1}^{(0)} < p_E$; otherwise, it stops.

Step (1): Step (1) commences with a fit of the logistic regression model containing x_{e_1}. Let $L_{e_1}^{(1)}$ denote the log-likelihood of this model. To determine whether any of the remaining $p - 1$ variables are important once the variable x_{e_1} is in the model, we fit the $p - 1$ logistic regression models containing x_{e_1} and x_j, $j = 1, 2, 3, ..., p$ and $j \neq e_1$. For the model containing x_{e_1} and x_j let the log-likelihood be denoted by $L_{e,j}^{(1)}$, and let the likelihood ratio chi-square statistic of this model versus the model containing only x_{e_1} be denoted by $G_j^{(1)} = 2(L_{e,j}^{(1)} - L_{e_1}^{(1)})$. The p-value for this statistic will be denoted by $p_j^{(1)}$. Let the variable with the smallest p-value at step (1) be x_{e_2} where $p_{e_2}^{(1)} = \min(p_j^{(1)})$. If this value is less than p_E we proceed to step (2); otherwise we stop.

Step (2): Step (2) begins with a fit of the model containing both x_{e_1} and x_{e_2}. It is possible that once x_{e_2} has been added to the model, x_{e_1} is no longer important. Thus, step (2) includes a check for backward elimination. In general this is accomplished by fitting models that delete one of the variables added in the previous steps and assessing the continued importance of the variable removed. At step (2) let $L_{-e_j}^{(2)}$ denote the log-likelihood of the model with x_{e_j} removed. In similar fashion let the likelihood ratio test of this model versus the full model at step (2) be $G_{-e_j}^{(2)} = 2(L_{e_1 e_2}^{(2)} - L_{-e_j}^{(2)})$ and $p_{-e_j}^{(2)}$ be its p-value.

To ascertain whether a variable should be deleted from the model the program selects that variable which, when removed, yields the maximum p-value. Denoting this variable as x_{r_2}, then $p_{r_2}^{(2)} = \max(p_{-e_1}^{(2)}, p_{-e_2}^{(2)})$. To decide whether x_{r_2} should be removed, the program compares $p_{r_2}^{(2)}$ to a second prechosen "alpha" level, p_R, which will indicate some minimal level of continued contribution to the model where "R" stands for remove. Whatever value we choose for p_R, it must exceed the value of p_E to guard against the possibility of having the program enter and remove the same variable at successive steps.

If we do not wish to exclude many variables once they have entered, we might use $p_R = 0.9$. A more stringent value would be used if a continued "significant" contribution were required. For example, if we used $p_E = 0.15$, then we might choose $p_R = 0.20$. If the maximum p-value to remove, $p_{r_2}^{(2)}$, exceeds p_R then x_{r_2} is removed from the model. If $p_{r_2}^{(2)}$ is less than p_R then x_{r_2} remains in the model. In either case the program proceeds to the variable selection phase.

At the forward selection phase each of the $p - 2$ logistic regression models are fit containing x_{e_1}, x_{e_2} and x_j for $j = 1, 2, 3, ..., p, j \neq e_1, e_2$. The program evaluates the log-likelihood for each model, computes the likelihood ratio test versus the model containing only x_{e_1} and x_{e_2} and determines the corresponding p-value. Let x_{e_3} denote the variable with the minimum p-value, that is,

$p_{e_3}^{(2)} = \min(p_j^{(2)})$. If this p-value is smaller than p_E, $p_{e_3}^{(2)} < p_E$, then the program proceeds to step (3); otherwise, it stops.

Step (3): The procedure for step (3) is identical to that of step (2). The program performs a check for backward elimination followed by forward selection. This process continues in this manner until the last step, step (S).

Step (S): This step occurs when: (1) all p variables have entered the model or (2) all variables in the model have p-values to remove which are less than p_R, and the variables not included in the model have p-values to enter which exceed p_E. The model at this step contains those variables that are important relative to the criteria of p_E and p_R. These may or may not be the variables reported in a final model. For instance, if the chosen values of p_E and p_R correspond to our belief for statistical significance, then the model at step S may well contain the significant variables. However, if we have used values for p_E and p_R which are less stringent, then we should select the variables for a final model from a table that summarizes the results of the stepwise procedure.

There are two methods that may be used to select variables from a summary table; these are comparable to methods commonly used in stepwise linear regression. The first method is based on the p-value for entry at each step, while the second is based on a likelihood ratio test of the model at the current step versus the model at the last step.

Let "q" denote an arbitrary step in the procedure. In the first method we compare $p_{e_q}^{(q-1)}$ to a prechosen significance level such as $\alpha = 0.15$. If the value $p_{e_q}^{(q-1)}$ is less than α, then we move to step q. We stop at the step when $p_{e_q}^{(q-1)}$ exceeds α. We consider the model at the previous step for further analysis. In this method the criterion for entry is based on a test of the significance of the coefficient for x_{e_q} conditional on x_{e_1}, x_{e_2}, ..., $x_{e_{q-1}}$ being in the model. The degrees of freedom for the test are 1 or $k - 1$, depending on whether x_{e_q} is continuous or polytomous with k categories.

In the second method, we compare the model at the current step, step q, not to the model at the previous step, step $q - 1$, but to the model at the last step, step (S). We evaluate the p-value for the likelihood ratio test of these two models and proceed in this fashion until this p-value exceeds α. This tests that the coefficients for the variables added to the model from step q to step (S) are all equal to zero. At any given step it will have more degrees of freedom than the test employed in the first method. For this reason the second method may possibly select a larger number of variables than the first method.

It is well known that the p-values calculated in stepwise selection procedures are not p-values in the traditional hypothesis testing context. Instead, they should be thought of as indicators of relative importance among variables. We recommend that one err in the direction of selecting a relatively rich model following stepwise selection. The variables so identified should then be subjected to the more intensive analysis described previously.

A common modification of the stepwise selection procedure just described is to begin with a model at step zero which contains known important covariates. Selection is then performed from among other variables. One instance when this approach may be useful is to select interactions from among those possible from a main effects model.

One considerable disadvantage of the stepwise selection procedures just described is that the maximum likelihood estimates for the coefficients of all variables not in the model must be calculated at each step. For large data files with large numbers of variables this can be quite costly both in terms of time and money. Two approximations to this method are available in, or could be implemented into, existing programs. One method, available in BMDP, uses a linear approximation to the likelihood ratio test. The resulting test is similar to the one used for variable selection in a two group stepwise discriminant analysis. This is termed the "ACE" method in BMDP. The second procedure selects new variables based on the Score tests for the variables not included in the model. A variant of this method using a multivariate Wald statistic has been proposed by Peduzzi, Hardy, and Holford (1980). To date there has been no work published which has compared these different selection methods although it does seem likely that an important variable would be identified, irrespective of method used.

Freedman (1983) urges caution when considering a model with many variables, noting that significant linear regressions may be obtained from variables completely unrelated to the outcome "noise" variable. Flack and Chang (1987) have shown similar results regarding the frequency of selection of "noise" variables. Thus, a thorough analysis that examines statistical and biologic significance is especially important following any stepwise method.

As an example, we apply the stepwise variable selection procedure to the low birth weight data. The results of this process are presented in Table 4.15 in terms of the p-values to enter and remove calculated at each step. These p-values are those of the relevant likelihood ratio test described previously. The order of the variables given columnwise in the table is the order in which they were selected. In each row the values to the left of the vertical line are p_R values and values to the right of the vertical lines are p_E values. The program was run using $p_E = 0.15$ and $p_R = 0.20$.

Step (0): At step (0) the program selects as a candidate for entry at step (1) the variable with the smallest p-value in the first row of Table 4.15. This is the variable PTL with a p-value of 0.009. Since this p-value is less than 0.15, the program proceeds to step (1).

Step (1): At step (1) the program will not remove the variable just entered since $p_R > p_E$ and the p-value to remove at step (1) is equal to the p-value to enter at step(0). (This is true for the variable entered at any step – not just the first step.) The variable with the smallest p-value to enter at step (1) is LWT with a value of 0.034, which is less than 0.15 so the program moves to step (2).

Step (2): The p-values to remove appear first in each row. The largest value is indicated with an "*." At step (2) the largest p-value to remove is 0.034, which does not exceed 0.20, thus the program moves to the variable selection phase. The smallest p-value to enter among the remaining variables not in the model is for the variable HT and is 0.006. This value is less than 0.15 so the program proceeds to step (3).

Step (3)–Step (5): At steps (3) to (5) the program finds that no variables can be removed from the model because each of the p-values, indicated with an "*" in rows three to five in Table 4.15, is less than 0.20. The program determines, in the selection phase, the variable with the smallest p-value to enter and, since it is less than 0.15, the program proceeds to the next step.

Step (6): At step (6) the program finds that the maximum p-value to remove is 0.135 for PTL. This value is less than 0.20, so PTL is not removed from the model. In the selection phase the program finds that the minimum p-

value for entry is 0.455 for the variable AGE. Since this value exceeds 0.15, no further variables may be entered into the model, and the program stops.

Since the program was run with $p_E = 0.15$, a value we believe will select variables with significant coefficients, it is not strictly necessary to go to the summary table to select the variables to be used in a final model. We will, however, illustrate the calculations for the two methods of variable selection from the summary table. These are given in Table 4.16.

Table 4.15 Results of Applying Stepwise Variable Selection Using the Maximum Likelihood Method to the Low Birth Weight Data Presented at Each Step in Terms of the p-Values to Enter, to the Right of the Vertical Line, and the p-Value Remove, to the Left of the Vertical Line in Each Row. The Asterisk Denotes the Maximum p-Value to Remove at Each Step.

Step #	PTL	LWT	HT	RACE	SMOKE	UI	AGE	FTV
0	0.009	0.015	0.045	0.082	0.027	0.024	0.097	0.379
1	0.009	0.034	0.038	0.069	0.078	0.083	0.057	0.441
2	0.022	0.034*	0.006	0.057	0.090	0.139	0.125	0.589
3	0.023*	0.006	0.006	0.078	0.093	0.086	0.162	0.731
4	0.019	0.005	0.009	0.078*	0.015	0.077	0.308	0.873
5	0.067*	0.009	0.010	0.016	0.015	0.088	0.374	0.905
6	0.135*	0.013	0.006	0.016	0.017	0.088	0.455	0.927

For method 1 we compare the p-value for entry at each step to our chosen level of significance. For purposes of illustration only we will use the value of 0.05, even though we noted earlier in this section that it is too stringent for actual practice. The information for method 1 is in the second panel of Table 4.16.

The value of the likelihood ratio test for the model at step (0) compared to that containing PTL at step (1) is

$$G = 6.78 = 2[-113.946 - (-117.336)]$$

The p-value for G is 0.009 which is less than 0.05 so we conclude that the coefficient for PTL is significant and move to step (2). The p-value for the variable, LWT, entered at step (2) is 0.034. This is the p-value for the likelihood ratio test of the significance of the coefficient for LWT, given that PTL is in the model. The value of the test statistic is

$$G = 4.49 = 2[-111.704 - (-113.946)]$$

Since the p-value for G is less than 0.05 we move to step (3). Calculations proceed in a similar fashion and we compare, at each step, the p-value to 0.05. At step (4) we find that the value of the likelihood ratio test of the model at step (4) versus that at step (3) is

$$G = 5.10 = 2[-105.43 - (-107.98)]$$

resulting in a p-value of 0.078. This value is greater than 0.05 so we conclude that RACE does not provide a significant addition to the variables already selected at step (3). Hence, the final model would be the one with all variables entered through step (3) even though the variable entered at step (5), SMOKE, has a p-value of less than 0.05.

Table 4.16 Log-Likelihood for the Model at Each Step and Likelihood Ratio Test Statistics (G), Degrees of Freedom (df), and p-Values for Two Methods of Selecting Variables for a Final Model from a Summary Table.

Variable			Method 1			Method 2		
Step #	Entered	Log-Likelihood	G	df	p-value	G	df	p-value
0		−117.34				32.69	7	<0.001
1	PTL	−113.95	6.78	1	0.009	25.91	6	<0.001
2	LWT	−111.70	4.48	1	0.034	21.42	5	0.001
3	HT	−107.98	7.44	1	0.006	13.98	4	0.007
4	RACE	−105.43	5.10	2	0.078	8.86	2	0.012
5	SMOKE	−102.45	5.95	1	0.015	2.91	1	0.088
6	UI	−100.99	2.91	1	0.088			

The information for method 2 is in the last panel of Table 4.15. In the second method the model at each step is compared to the model at the last step via a likelihood ratio test. This is a test of the joint significance of variables added at subsequent steps. We again proceed until the p-value for the test exceeds the chosen significance level. For purposes of illustration only we will use 0.05. The value of G at step (0) is

$$G = 2[-100.993 - (-117.336)] = 32.69$$

with a p-value of <0.001 based on 7 degrees of freedom. Since this p-value is less than 0.05 we proceed to step (1). At step (1) the test of this model versus that at the last step is

$$G = 2[-100.993 - (-113.946)] = 25.91$$

with a p-value of <0.001 based on 6 degrees of freedom. Since the p-value is less than 0.05 we proceed to step (3). We continue in this manner until step (5). The p-value for the likelihood ratio test of the model at step (5) versus that at step (6) is 0.088. This value exceeds 0.05, so we stop and use the variables in the model at step (5).

In this example methods 1 and 2 have identified different sets of variables. Each method provides a test of a different hypothesis at each step. The number of parameters being tested in method 2 is, except for the last step, larger than that for method 1. Thus, method 2 may select, as it does in this example, more variables than method 1. In cases where this occurs, one should carefully examine the additional variables and include them if they seem biologically relevant. In this case we would undoubtably opt for the richer model selected by method 2.

At the conclusion of the stepwise selection process we have only identified a collection of variables which seem to be statistically important. Thus, any known biologically important variables, such as AGE in our example, should be added before proceeding with the steps necessary to obtain the final main effects model. As noted earlier, this should include determining the appropriate scale of continuous covariates.

Once the scale of the continuous covariates has been examined, and corrected if necessary, we may consider applying stepwise selection to identify interactions. The candidate interaction terms are those that seem biologically reasonable given the main effects variables in the model. We begin at step (0) with the main effects model and sequentially select from among the possible interactions. We select the significant ones using either method 1 or method 2. Consequently the final model will contain previously identified main effects and significant interaction terms.

The variables identified by the stepwise selection process in the low birth weight data are the same ones identified earlier by purposeful selection. Therefore, the work necessary to check the scale of continuous covariates is not repeated and we begin stepwise selection of interactions using the model given in

Table 4.8 and the interactions listed in Table 4.9. The results of stepwise selection of interactions are given in Table 4.17.

Table 4.17 Results of Applying Stepwise Variable Selection to Interactions from the Main Effects Model, Using the Maximum Likelihood Method Presented at Each Step in Terms of the *p*-Values to Enter, to the Right of the Vertical Line, and the *p*-Values to Remove to the Left of the Vertical Line. The Asterisk Denotes the Maximum *p*-Value to Remove at Each Step.

Step #	AGE ×LWD	RACE ×LWD	HT×LWD	SMOKE ×LWD
0	0.110	0.183	0.291	0.127
1	0.110*	0.081	0.252	0.084
2	0.041	0.081*	0.142	0.645
3	0.029	0.053	0.142*	0.562

Of the 16 possible interactions specified in Table 4.9, only three were chosen. After step (3) the model was overfit, a point we discuss in Section 4.5. In the last column of Table 4.17 we have given the *p*-value for entering the SMOKE × LWD interaction term, which was included in the model given in Table 4.10. The results at step (1) indicate that the RACE × LWD interaction is negligibly more significant than the SMOKE × LWD interaction and, once the RACE × LWD interaction is included into the model, there is little additional importance in the SMOKE × LWD interaction. At step (3) we see that the HT × LWD interaction enters the model with a *p*-value of 0.142.

We now face several decisions involving the interactions. We considered this same problem earlier in this chapter when discussing purposeful selection of variables. For the sake of completeness, we repeat the analysis in the current context. To further explore the tradeoff between including the SMOKE × LWD or the RACE × LWD interaction, a model that forced the SMOKE × LWD interaction into the model and then added the RACE × LWD interaction was fit. The results showed, as expected, that the RACE × LWD interaction was no longer important once the SMOKE × LWD interaction was included in the model. We must, therefore, decide which of these two interactions to include. We choose the SMOKE × LWD interaction as the more important from the biologic standpoint in view of the known relationship between weight and smoking. Potential racial by weight differences are regarded as being of lesser importance to document.

We now must decide if the HT × LWD interaction should be added to the model. The *p*-value for the inclusion of this interaction after the SMOKE × LWD interaction term is included in the model is 0.160, again a value close to the preferred alpha of 0.15. At this point we must keep in mind that the fundamental reason for developing a model is to provide as clear a description as is possible with the available data of the associations between outcome and covariates. If entering an additional term into the model improves our estimates of the relevant associations then we should put that term into the model regardless of its statistical significance. If a term does not contribute to the overall goal then it may be excluded. In this case we determine that inclusion of the HT × LWD interaction term does not help our understanding of the association between low birth weight and the variables in the model so we choose to leave it out of the model. The resulting model is the same as the one given in Table 4.10. It is unnecessary to repeat the computation of estimated odds ratios. Had we used the model containing three interactions, then estimated odds ratios similar to those shown in Table 4.13 would be calculated. In this case the odds ratios for each of LWD, SMOKE, and HT would be computed at each level of the combination of the other two variables leading to four estimated odds ratios per effect.

We noted that approximations to selection by likelihood ratio tests are available which are computationally much faster and generally yield identical variable selection. To provide a comparison we present the *p*-values for entry and removal using the approximation method available in BMDPLR in Table 4.18. The values are quite close to those shown in Table 4.15. Use of the *p*-values in Table 4.18 would select the same variables as those selected from Table 4.15. The reversal in the order of selection of LWT and HT does not effect their inclusion in the list for a final model. Since stepwise selection should serve primarily as a guide to us for variable selection there is little reason to prefer the more intensive maximum likelihood method in practice. Computational issues will become less important in the future as faster microcomputers and good software are developed.

Table 4.18 Results of Applying Stepwise Variable Selection Using the Approximation Method Presented at Each Step in Terms of the p-Values to Enter, Above the Line in Each Column and the p-Value Remove, Below the Line in Each Column.

Step #	PTL	HT	LWT	RACE	SMOKE	UI	AGE	FTV
0	0.007	0.036	0.019	0.082	0.027	0.020	0.103	0.389
1	0.013*	0.030	0.043	0.070	0.077	0.077	0.064	0.451
2	0.011	0.042*	0.008	0.089	0.079	0.041	0.065	0.548
3	0.032*	0.011	0.013	0.080	0.098	0.086	0.179	0.738
4	0.026	0.014	0.011	0.088*	0.016	0.076	0.322	0.876
5	0.077*	0.014	0.018	0.024	0.019	0.085	0.383	0.906
6	0.145*	0.009	0.023	0.022	0.021	0.089	0.462	0.815

In conclusion, stepwise selection identifies variables as candidates for a model solely on statistical grounds. Thus, following stepwise selection of main effects all variables should be carefully scrutinized for biologic plausibility. In general, interactions must attain at least a moderate level of statistical significance to alter the point and interval estimates from a main effects model. Thus, stepwise selection of interactions can provide a valuable contribution to model identification, especially when there are large numbers of biologically plausible interactions generated from the main effects.

4.4 Best Subsets Logistic Regression

An alternative to stepwise selection of variables for a model is best subsets selection. This approach to model building has been available for linear regression for a number of years and makes use of the branch and bound algorithm of Furnival and Wilson (1974). Typical software implementing this method for linear regression will identify a specified number of "best" models containing one, two, three variables, and so on, up to the single model containing all p variables. Lawless and Singhal (1978, 1987a, 1987b) proposed an extension that may be used with any non-normal model. The crux of their method involves application of the Furnival–Wilson algorithm to a linear approximation of the cross-product sum-of-squares matrix, which yields approximations to the maximum likelihood estimates. Selected models are then compared to the model containing all variables via a likelihood ratio test. Hosmer, Jovanovic, and Lemeshow (1989) have shown that in the case of logistic regression the full generality of the Lawless and Singhal approach is not needed. Best subsets logistic regression may be performed in a straightforward manner using any program for best subsets linear regression.

Application of best subsets linear regression software to perform best subsets logistic regression is most easily explained using vector and matrix notation. In this regard we let $\mathbf{X}$ denote the $n \times (p + 1)$ matrix containing the values of all p independent variables for each subject, with the first column containing 1 to represent the constant term. Here the p variables may represent the total number of variables, or those selected at the univariate stage of model building. We let $\mathbf{V}$ denote an $n \times n$ diagonal matrix with general element $v_i = \hat{\pi}_i(1 - \hat{\pi}_i)$ where $\hat{\pi}_i$ is the estimated logistic probability computed using the maximum likelihood estimate $\hat{\boldsymbol{\beta}}$ and the data for the ith case, $\mathbf{x}_i$.

For sake of clarity of presentation in this section we repeat the expression for $\mathbf{X}$ and $\mathbf{V}$ given in Chapter 2. They are as follows:

$$\mathbf{X} = \begin{bmatrix} 1 & x_{11} & \cdots & x_{1p} \\ 1 & x_{21} & \cdots & x_{2p} \\ & \vdots & & \\ 1 & x_{n1} & \cdots & x_{np} \end{bmatrix}$$

and

$$\mathbf{V} = \begin{bmatrix} \hat{\pi}_1(1 - \hat{\pi}_1) & 0 & \cdots & 0 \\ 0 & \hat{\pi}_2(1 - \hat{\pi}_2) & \cdots & 0 \\ & & \vdots & \\ 0 & 0 & \cdots & \hat{\pi}_n(1 - \hat{\pi}_n) \end{bmatrix}$$

As noted in Chapter 2, the maximum likelihood estimate is determined iteratively. It may be shown [see Pregibon (1981)] that $\hat{\boldsymbol{\beta}} = (\mathbf{X}'\mathbf{V}\mathbf{X})^{-1}\mathbf{X}'\mathbf{V}\mathbf{z}$, where the vector $\mathbf{z}$ contains pseudovalues, $\mathbf{z} = \mathbf{X}\hat{\boldsymbol{\beta}} + \mathbf{V}^{-1}\mathbf{r}$ and $\mathbf{r}$ is the vector of residuals, $\mathbf{r} = (\mathbf{y} - \hat{\boldsymbol{\pi}})$. This representation of $\hat{\boldsymbol{\beta}}$ provides the basis for use of linear regression software. It is easy to verify that any linear regression package allowing weights will produce coefficient estimates identical to $\hat{\boldsymbol{\beta}}$ when used with z as the dependent variable and case weights, v_i, equal to the diagonal elements of $\mathbf{V}$.

To replicate the results of the maximum likelihood fit from a logistic regression package using a linear regression package we would calculate for each case the value of a pseudodependent variable

$$z_i = (1, x_i)\, \hat{\boldsymbol{\beta}} + \frac{(y_i - \hat{\pi}_i)}{\hat{\pi}_i(1 - \hat{\pi}_i)}$$

$$= \hat{\beta}_0 + \sum_{j=1}^{p} \hat{\beta}_j x_j + \frac{(y_i - \hat{\pi}_i)}{\hat{\pi}_i(1 - \hat{\pi}_i)}$$

$$= \ln\left(\frac{\hat{\pi}_i}{1 - \hat{\pi}_i}\right) + \frac{(y_i - \hat{\pi}_i)}{\hat{\pi}_i(1 - \hat{\pi}_i)} \tag{4.7}$$

and a case weight

$$v_i = \hat{\pi}_i(1 - \hat{\pi}_i) \tag{4.8}$$

Note that all we need is access to the fitted values, $\hat{\pi}_i$, to compute the values of z_i and v_i. Next we would run the linear regression program using the values of z_i for our dependent variable, the values of x_i for our vector of independent variables, and the values of v_i for our case weights.

Proceeding further with the linear regression, it can be shown that the residuals from the fit are

$$(z_i - \hat{z}_i) = \frac{(y_i - \hat{\pi}_i)}{\hat{\pi}_i(1 - \hat{\pi}_i)}$$

and the weighted residual sum-of-squares produced by the program is

$$\sum_{i=1}^{n} v_i(z_i - \hat{z}_i)^2 = \sum_{i=1}^{n} \frac{(y_i - \hat{\pi}_i)^2}{\hat{\pi}_i(1 - \hat{\pi}_i)}$$

which is X^2, the Pearson chi-square statistic from a maximum likelihood logistic regression program. It follows that the mean residual sum-of-squares is $s^2 = X^2/(n - p - 1)$. The estimates of the standard error of the estimated coefficients produced by the linear regression program are s times the square root of the diagonal elements of the matrix $(\mathbf{X'VX})^{-1}$. Thus, to obtain the correct values given in equation (2.5) we need to divide the estimates of the standard error produced by the linear regression program by s, the square root of the mean square error (or standard error of the estimate).

The ability to duplicate the maximum likelihood fit in a linear regression package forms the foundation of the suggested method for performing best subsets logistic regression. In particular, Hosmer, Jovanovic, and Lemeshow (1989) show that use of any best subsets linear regression program with values of z in equation (4.7) for the dependent variable, case weights v_i shown in equation (4.8), and covariates $\mathbf{x}$, produces for any subset of q variables the approximate coefficient estimates of Lawless and Singhal (1978). Hence, we may use any best subsets linear regression program to execute the computations for best subsets logistic regression.

The subsets of variables selected for "best" models will depend on the criterion chosen for "best." In best subsets linear regression three criteria have primarily been used to select variables. Two of these are based on the concept of the proportion of the total variation explained by the model. These are R^2, the ratio of the regression sum-of-squares to the total sum-of-squares, and adjusted R^2 or AR^2, the ratio of the regression mean squares to the total mean squares. Since the adjusted R^2 is based on mean squares rather than sums-of-squares, it provides a correction for the number of variables in the model. This is important as we must be able to compare models containing different variables and different numbers of variables. If we use R^2, the best model is always the model containing all p variables, a result that is not very helpful. An obvious extension for best subsets logistic regression is to base the R^2 measures, in a manner similar to that shown in Chapter 5, on deviance rather than Pearson chi-square. However, we do not recommend the use of the R^2 measures for best subsets logistic regression. Instead, we prefer to use the third measure used in best subsets linear regression that was developed by Mallows (1973). This is a measure of predictive squared error, denoted C_q. (This measure is denoted as C_p by other authors. We use "q" instead of "p" in this text since the letter p refers to a total number of possible variables while q refers to some subset of variables.)

A summary of the development of criterion C_q follows. Let $\gamma(x_i)$ be the true value for the conditional mean, $E(Y_i \mid x_i)$, and let $\hat{y}_i$ be the fitted value under the hypothesized model containing q of the p possible variables. The scaled predictive squared error (PE_q) for the fitted model is,

$$PE_q = \sum [\hat{y}_i - \gamma(x_i)]^2 / \sigma^2$$

where σ^2 is the error variance under the model. Under the usual normal errors linear regression model the expected value of PE_q is

$$E(PE_q) = (q + 1) + SSB_q / \sigma^2$$

where the sum of squares between models, SSB_q, is $SSB_q = \Sigma[\eta(x_i) - \gamma(x_i)]^2$, and $\eta(x_i)$ is the conditional mean under the hypothesized model. For example, if the true model is a linear regression in two variables and the hypothesized model is a linear regression in one of these variable then

$$\gamma(x_i) = \beta_0 + \beta_1 x_{1i} + \beta_2 x_{2i}$$

and

$$\eta(x_i) = \xi_0 + \xi_1 x_{1i}$$

When the hypothesized model being fit is correct we have $\eta(x_i) = \gamma(x_i)$ for each i and $SSB_q = 0$, and we expect PE_q to be equal to $(q + 1)$, the number of parameters in the model. Mallow's C_q is obtained by estimating the quantities in the expression for $E(PE_q)$.

Under the assumption of a linear regression with normal errors model, the expected value for the residual sum-of-squares (RSS_q) is

$$E(RSS_q) = (n - q + 1)\sigma^2 + SSB_q$$

We estimate the value of σ^2 by the residual mean square from the model containing all p variables, denoted by $\hat{\sigma}^2$. We estimate $E(\text{RSS}_q)$ by its observed value RSS_q. Substituting these into the expression for $E(\text{PE}_q)$ yields

$$C_q = \frac{\text{RSS}_q}{\hat{\sigma}^2} - n + 2(q + 1)$$

When we use C_q in linear regression the number of parameters in the model provides us with a reference standard. In particular, if C_q is less than $q + 1$ then the fitted model has less error than the model containing all p variables. When the value of C_q exceeds $q + 1$ then the fitted model contains more error than the model with all p variables. The measure is created in such a way that its value when the subset contains all p variables is $C_p = p + 1$. Linear regression packages choose as the best model the one with the smallest value of C_q.

Hosmer, Jovanovic, and Lemeshow (1989) show that when best subsets logistic regression is performed via a best subsets linear regression package in the manner described previously in this section, Mollow's C_q has the same intuitive appeal. In particular they show that for a subset of q of the p variables

$$C_q = \frac{X^2 + \lambda^*}{X^2/(n - p - 1)} + 2(q + 1) - n$$

where $X^2 = \Sigma\{(y_i - \hat{\pi}_i)^2/[\hat{\pi}_i(1 - \hat{\pi}_i)]\}$, the Pearson chi-square statistic for the model with p variables and λ^* is the multivariate Wald test statistic for the hypothesis that the coefficients for the $p - q$ variables not in the model are equal to zero. Under the assumption that the model fit is the correct one, the approximate expected values of X^2 and λ^* are $(n - p - 1)$ and $p - q$ respectively. Substitution of these approximate expected values into the expression for C_q yields $C_q = q + 1$. Hence, models with C_q near $q + 1$ will be candidates for a best model. The best subsets linear regression program will select as best that subset with the smallest value of C_q.

Use of the best subsets linear regression package should help select, in the same way its application in linear regression does, a core of q important covariates from the p possible covariates. At this point, we suggest that further modeling proceed in the manner described for purposeful selection of variables using a logistic regression package. Users should not be lured into accepting the variables suggested by a best subset strategy without considerable critical evaluation.

We illustrate "best" subsets selection with the low birth weight data. The variables used were those indicated in Table 4.1, except the number of third trimester physician visits. All variables except RACE were treated as if they were continuous. RACE was coded into the two design variables shown in previous sections. A logistic regression package was used to obtain the estimated logistic probabilities for the model containing all $p = 8$ variables. Following the fit of the full model the values of z and v were created using equations (4.7) and (4.8). A best subsets linear regression package was used with z as the dependent variable and v as the case weights. The possible independent variables were the 7 continuous variables plus two design variables for RACE, for a total of 9. The best subsets linear regression program used did not have a provision for the creation of design variables from categorical scaled covariates so the design variables for RACE were created prior to the best subsets analysis.

In Table 4.19 we present the results of the five best models selected using C_q as the criterion. For comparative purposes we also present the maximum likelihood estimates of the coefficients obtained from a logistic regression package. In the bottom panel of the Table 4.19 are various summary statistics resulting from the weighted linear regression least squares fit and the maximum likelihood fit. These include the values of C_q, the value of λ^*, and the value of the likelihood ratio test, G, for the variables excluded from the model. The test statistics G and λ^* have similar values since they test the same hypothesis and have the same asymptotic distribution.

Looking first at the estimates of the coefficients, we see that for each of the five models the weighted least squares linear regression estimates are remarkably close to the maximum likelihood estimates. This observation should not be taken as an endorsement for using weighted least squares with pseudodata in place of maximum likelihood, but rather as support of the adequacy of its use for variable selection purposes. Since best subsets selection identifies models whose coefficients are similar to those from the full model we might suspect that the selection process would favor a richer model. However, variables in the full model whose coefficients are "zero" should not force the selection in the direction of models with more variables. In this example, the "best" models identified are in close agreement with those identified by both purposeful and stepwise selection. The advantage of best subsets selection is that we have been provided with an in-depth look at a much greater number of models.

Table 4.19 Weighted Least Squares Estimates, Maximum Likelihood Estimates, and Summary Statistics for the Best Five Models Selected Using Best Subsets Linear Regression with Pseudodata and Mallow's C_q as the Criterion.

Variable	MLE	WLS	MLE	WLS	MLE	WLS	MLE	WLS	MLE	WLS
AGE							-0.03	-0.03	-0.02	-0.02
RACE (1)	1.32	1.32	1.32	1.31	1.29	1.27	1.26	1.26	1.28	1.27
RACE (2)	0.89	0.89	0.93	0.91	0.91	0.89	0.86	0.86	0.90	0.89
SMOKE	0.94	0.93	1.04	1.01	0.95	0.92	0.92	0.92	1.03	1.01
HT	1.86	1.84	1.87	1.85	1.74	1.72	1.83	1.83	1.86	1.83
UI	0.79	0.78	0.90	0.89			0.76	0.75	0.89	0.88
LWT	-0.02	-0.02	-0.02	-0.02	-0.02	-0.02	-0.02	-0.02	-0.02	-0.02
PTL	0.50	0.50			0.60	0.59	0.54	0.54		
Constant	-0.09	-0.08	0.06	0.05	0.11	0.11	0.46	0.46	0.43	0.42
Summary										
C_q		6.68		6.79		7.54		8.14		8.54
λ^*		0.7		2.8		3.6		0.1		2.59
G	0.7		2.9		3.6		0.1		2.7	

The summary statistics from the weighted least squares and maximum likelihood fits are also quite similar. The values of λ^* and G are in close agreement. The degrees of freedom for these tests are equal to the number of variables with coefficients set equal to zero (blank in Table 4.19) plus 1 to account for the exclusion of FTV from all models listed.

Using only the summary statistics, we would select model 1 as the best model. However, this model does not contain AGE, which is an important biologic variable, and which may be needed for assessment of interactions. Hence, we recommend including AGE and proceed to the next stage of model development using model 4. At this point model building is identical to that already presented in Section 4.1 where we focused on scale identification of continuous variables. Once we have finalized the main effects model, we could employ best subsets selection to decide on possible interactions.

The advantage of the proposed method of best subsets logistic regression is that many more models can be quickly screened than was possible with the other approaches to variable identification. There is, however, one potential disadvantage with the best subsets approach: we must be able to fit a full model. In analyses that include a large number of variables this may not be possible. Numerical problems can occur when we overfit a logistic regression model. These are discussed in the next section; but suffice it to say that if the model has many variables, then we run the risk that data will be too thin to be able to estimate all the parameters. If the full model proves to be too rich, then some selective weeding out of obviously unimportant variables with univariate tests may remedy this problem. Another approach is to perform the best subsets analysis using several smaller "full" models.

In summary, the ability to use weighted least squares best subsets linear regression software to identify variables for logistic regression should be kept in mind as a possible aid to variable selection. As is the case with any statistical selection method, the biologic basis of all variables should be addressed before any model is accepted as the final model.

4.5 Numerical Problems

In previous chapters we have occasionally mentioned various numerical problems that can occur when fitting a logistic regression model. These problems are caused by certain structures in the data and the lack of appropriate checks in logistic regression software. The goal of this section is to illustrate these structures in certain simple situations and illustrate what can happen when the logistic regression model is fit to such data. The issue here is not one of model correctness or specification, but the effect certain data patterns have on the computation of parameter estimates.

Perhaps the simplest and thus most obvious situation is when we have a frequency of zero in a contingency table. An example of such a contingency table is given in Table 4.20. The estimated odds ratios and log odds ratios using the first level of the covariate as the reference group are given in the first two rows below the table. The point estimate of the odds ratios for the level 3 versus level 1 is infinite since all subjects at level 3 responded. The results of fitting a logistic regression model to these data are given in the last two rows. The estimated coefficient in the first column is the intercept coefficient. The particular package used does not really matter as most packages will produce similar output. The program may or may not provide some sort of error message indicating that convergence was not obtained or that the maximum number of iterations was used. What is rather obvious, and the tip-off that there is a problem with the model, is the large estimated coefficient for the second design variable and especially its large estimated standard error.

Table 4.20   A Contingency Table with a Zero Cell Count and the Results of Fitting a Logistic Regression Model to This Data.

Outcome	1	2	3	Total
1	7	12	220	39
0	13	8	0	21
Total	20	20	20	60
$\hat{\psi}$	1	2.79	inf	
$\ln(\hat{\psi})$	0	1.03	inf	
$\hat{\beta}$	−0.62	1.03	11.7	
SE	0.47	0.65	34.9	

A common practice to avoid having an undefined point estimate is to add one-half to each of the cell counts. Adding one-half may allow us to move forward with the analysis of a single contingency table but such a simplistic remedy will rarely be satisfactory with a more complex data set.

As a slightly more complex example we consider the stratified 2 by 2 tables shown in Table 4.21. The stratum-specific point estimates of the odds ratios are provided below each 2 by 2 table. The results of fitting a series of logistic regression models are provided in Table 4.22.

Table 4.21   Stratified 2 by 2 Contingency Tables with a Zero Cell Count Within One Strata.

Stratum	1		2		3	
Outcome	1	0	1	0	1	0
1	5	2	10	2	15	1
0	5	8	2	6	0	4
Total	10	10	12	8	15	5
$\hat{\psi}$	4		15		inf	

In the case of the data shown in Table 4.21 we do not encounter problems until we add the stratum, z, by risk factor, x, interaction terms, $x \times z$ into the model. The addition of the interaction term results in a model that is equivalent to fitting a model with a single categorical variable with six levels, one for each column in Table 4.21. Thus, in a sense the problem encountered when we include the interaction is the same one illustrated in Table 4.20. As was the case when fitting a model to the data in Table 4.20 the presence of a zero cell count is manifested by an unbelievable large estimated coefficient and estimated standard error.

Table 4.22   Results of Fitting Logistic Regression Models to the Data in Table 8.13.

Model	1		2	
Variable	Est. Coeff.	Est. SE	Est. Coeff.	Est. SE
x	2.77	0.72	1.39	1.01
$z(1)$	1.19	0.81	0.29	1.14
$z(2)$	2.04	0.89	0	1.37
$x \times z(1)$			1.32	1.51
$x \times z(2)$			11.54	50.22
Constant	−2.32	0.77	−1.39	0.79

The presence of a zero cell count should be detected during the univariate screening of the data. Knowing that the zero cell count is going to cause problems in the modeling stage of the analysis we could collapse the categories of the variable in a meaningful way to eliminate it, eliminate the category all together or, if the variable is at least ordinal scaled, treat it as continuous.

The type of zero cell count illustrated in Table 4.21 results from spreading our data over too many cells. This problem will likely not occur until we begin to include interactions into the model. When it does occur we should examine the three way contingency table equivalent to the one shown in Table 4.21. The unstable results prevents us from ascertaining if in fact the interaction is important. To assess the interaction we first need to eliminate the zero cell count. One way to do this is by collapsing categories of the stratification variable. For example, in Table 4.21 we might decide that values of $z = 2$ and $z = 3$ are similar enough to pool them. The stratified analysis would then have two 2 by 2 tables the second of which results from pooling the tables for $z = 2$ and $z = 3$. A second approach is to define a new variable equal to the combination of the stratification variable and the risk factor and to pool over levels of this variable and model it as a main effect variable. Using Table 4.21 as an example we would have a variable with six levels corresponding to the six columns in the table. We could collapse levels five and six together. Another pooling strategy would be to pool levels three and five and four and six. This pooling strategy is equivalent to collapsing over levels of the stratification variable. The net effect is the loss of degrees of freedom commensurate with the amount of pooling. Twice the difference in the log-likelihood for the main effects only model and the model with the modified interaction term added provides a statistic for the significance of the coefficients for the modified interaction term.

The fitted models shown in Tables 4.20 and 4.22 resulted in very large estimated coefficients and estimated standard errors. In some examples we have encountered the magnitude of the estimated coefficient was not large enough to suspect a numerical problem; but the estimated standard error always was. Hence, we believe that the best indicator of numerical problems in logistic regression is the estimated standard error. In general anytime the estimated standard error of an estimated coefficient is very large relative to the point estimate we should suspect the presence of one of the data structures described in this section.

A second type of numerical problem occurs when a collection of the covariates completely separates the outcome groups or, in the terminology of discriminant analysis, the covariates discriminate perfectly. For example, suppose that the age of every subject with the outcome present was greater than 50 and the age of all subjects with the outcome absent was less than 49. Thus, if we know the age of a subject we know with certainty the value of the outcome variable. In this situation there is no overlap in the distribution of the covariates between the two outcome groups. This type of data has been shown by Bryson and Johnson (1981) to have the property of monotone likelihood. The net results of this is that the maximum likelihood estimates do not exist [Albert and Anderson (1984) and Santner and Duffy (1986)]. In order to have finite maximum likelihood estimates we must have some overlap in the distribution of the covariates in the model.

A simple example will illustrate the problem of complete separation and the results of fitting logistic regression models to such data. Suppose we have the following 12 pairs of covariate and outcome, (x,y): (1,0), (2,0), (3,0), (4,0), (5,0), ($x = 5.5$, or 6.0, or 6.05, or 6.1, or 6.2, or 8.0, $y = 0$), (6,1), (7,1), (8,1), (9,1), (10,1), (11,1). The results of fitting logistic regression models and letting x for case 6 take on one of the values 5.5, 6.0, 6.05, 6.1, 6.2, or 8, are given in Table 4.23. When we use $x = 5.5$ we have complete separation and all estimated parameters are huge, since the maximum likelihood estimates do not exist. Similar behavior occurs when the value of $x = 6.0$ is used. When overlap is at a single or a few tied values the configuration was termed by Albert and Anderson as quasicomplete separation. As the value of x takes on values greater than 6 the overlap becomes greater and the estimated parameters and standard errors begin to attain more reasonable values. The sensitivity of the fit to the overlap will of course depend on the sample size and the range of the covariate. The tip-off that something is amiss is, as in the case of the zero cell count, the very large estimated coefficients and especially the large estimated standard errors.

The occurrence of complete separation in practice will depend on the sample size, the number of subjects with the outcome present, and the number of variables included into the model. For example, suppose we have a sample of 25 subjects and only five have the outcome present. The chance that the main effects model will demonstrate complete separation will increase with the number of variables we include in the model. Thus, the modeling strategy that includes all variables into the model will be particularly sensitive to complete separation. Albert and Anderson and Santner and Duffy provide rather complicated diagnostic procedures for determining whether a set of data displays complete or quasicomplete separation. Albert and Anderson recommend that in the absence of their diagnostic, one look at the estimated standard errors and if these tend to increase substantially with each iteration of the fit, that one suspect the presence of complete separation. As noted in Chapter 3 the easiest way to address complete separation (it was called over fitting in Chapter 3) is to use some careful univariate analyses. The occurrence of complete separation is not likely to be of great biologic importance as it is usually a numerical coincidence rather than describing some important biologic phenomenon. It is a problem we will have to work around.

Table 4.23 Estimated Slope (x), Constant, and Estimated Standard Errors When the Data Have Complete Separation, Quasicomplete Separation, and Overlap.

Estimates	5.5	6	6.05	6.1	6.15	6.2	8.0
x	20.6	7.2	4.3	3.6	3.2	2.9	1.0
SE	22.7	15.8	6.1	4.1	3.3	2.8	0.5
Constant	−118.3	−43.3	−26.2	−21.9	−19.5	−17.8	−6.1
SE	130.2	95.0	36.6	25.1	20.3	16.9	3.5

As is the case in linear regression, model fitting via logistic regression is also sensitive to colinearities among the independent variables in the model. Most software packages will have some sort of diagnostic check like the tolerance test employed in linear regression. Nevertheless it is possible for variables to pass these tests and have the program run but yield output that is clearly nonsense. As a simple example we fit logistic regression models to the data displayed in Table 4.24. In the table $x_1 \sim N(0,1)$ and the outcome variable was generated by comparing a $U(0,1)$ variate, u, to the true probability $\pi(x_1) = e^{x_1}/(1 + e^{x_1})$ as follows: if $u < \pi(x_1)$ then $y = 1$, otherwise $y = 0$. The notation $N(0,1)$ indicates a random variable following the standard normal (mean = zero, variance = 1) distribution and $U(a,b)$ indicates a random variable following the uniform distribution on the interval $[a,b]$. The other variables were generated from x_1 and the constant as follows: $x_2 = x_1 + U(0,0.01)$ and $x_3 = 1 + U(0,0.01)$. Thus, x_1 and x_2 are highly correlated and x_3 is nearly colinear with the constant term. The results of fitting logistic regression models to various subsets of the variables shown in Table 4.24 are presented in Table 4.25.

Table 4.24 Data Displaying Near Colinearity Among the Independent Variables and Constant.

Subject	x_1	x_2	x_3	y
1	0.225	0.231	1.026	0
2	0.487	0.489	1.022	1
3	−1.080	−1.070	1.074	0
4	−0.870	−0.870	1.091	0
5	−0.580	−0.570	1.095	0
6	−0.640	−0.640	1.010	0
7	1.614	1.619	1.087	0
8	0.352	0.355	1.095	1
9	−1.025	−1.018	1.008	0
10	0.929	0.937	1.057	1

The model that includes the highly correlated variables x_1 and x_2 has both very large estimated slope coefficients and estimated standard errors. For the model containing x_3 we see that the estimated coefficients are of reasonable magnitude but the estimated standard errors are much larger than we would expect. The model containing all variables is a composite of the results of the other models. In all cases the tip-off for a problem comes from the aberrantly large estimated standard errors.

In a more complicated data set an analysis of the associations among the covariates using a colinearity analysis similar to that performed in linear regression should be helpful in identifying the dependencies among the covariates [see Belsley, Kuh, and Welsch (1980)]. One would normally not employ such an in-depth investigation of the covariates unless there was evidence of degradation in the fit similar to that shown in Table 4.25. An alternative is to use the ridge regression methods proposed by Schaefer (1986).

Table 4.25 Estimated Coefficients and Standard Errors Resulting From Fitting Logistic Regression Models to the Data in Table 4.24.

Variable	Coeff.	SE	Coeff.	SE	Coeff.	SE	Coeff.	SE
x_1	1.4	1.00	502.1	444.3			500.2	446.5
x_2			−500.4	443.8			−498.6	446.0
x_3					1.59	19.89	−1.3	28.5
Constant	−1.0	0.83	1.28	1.98	−2.53	21.04	2.7	29.8

In general the numerical problems of a zero cell count, complete separation, and colinearity, are always manifested by extraordinary large estimated standard errors and sometimes by a large estimated coefficient as well. New users and ones without much computer experience are especially cautioned to look at their results carefully for evidence of numerical problems. Consultation with someone more experienced may be required to ferret out and solve these numeric problems.

Exercises

1. Selection of the scale for continuous covariates is an important step in any modeling process. The variable systolic blood pressure at admission, SYS, in the ICU study described in Section 1.5 presents a particularly challenging example. Consider the variable vital status (STA) as the outcome variable and SYS as the covariate for a univariate logistic regression model. What is the correct scale for SYS to enter into the model? As as second example consider a univariate model with heart rate at ICU admission (HRA) as the covariate. Repeat this exercise of scale identification for SYS and HRA using a multivariate model containing these two variables plus three or four other covariates of your choice. If the parametric form of the logit appears to be overly complicated, involving higher order terms like the cube and fourth power of the variable, then an alternative is to use design variables based on the quartiles or quintiles of the variable. Is this a reasonable approach to modeling SYS?

2. Consider the variable level of consciousness at ICU admission (LOC) as a covariate and vital status (STA) as the outcome variable. Compare the

estimates of the odds ratios obtained from the cross-classification of STA by LOC and the logistic regression of STA on LOC. Use LOC = 0 as the reference group for both methods. How well did the logistic regression program deal with the zero cell? What strategy would you adopt to modeling LOC in future analyses?

3.   Consider the variable vital status (STA) as the outcome variable and the remainder of the variables in the ICU data set as potential covariates. Use each of the variable selection methods discussed in this chapter to find a "best" model. Document thoroughly the rationale for each step in the process you follow. Compare and contrast the models resulting from the different approaches to variable selection. Note that in all cases the analysis should address not only identification of main effects but also appropriate scale for continuous covariates and potential interactions. Display the results of your final model in a table. Include in the table point and 95% CI estimates of all relevant odds ratios.

4.   Repeat problems 1 and 3 for as many other data sets as you have access to. In addition try different software packages and note their strengths and weaknesses for each particular method.

CHAPTER 5

Assessing the Fit of the Model

5.1 Introduction

We begin our discussion of methods for assessing the fit of an estimated logistic regression model with the assumption that we are at least preliminarily satisfied with our efforts at the model building stage. By this we mean that, to the best of our knowledge, the model contains those variables (main effects as well as interactions) that should be in the model and that variables have been entered in the correct functional form. Now we would like to know how effective the model we have is in describing the outcome variable. This is referred to as its **goodness-of-fit**.

If we intend to assess the goodness-of-fit of the model, then we should have some specific ideas about what it means to say that a model fits. Suppose we denote the observed sample values of the outcome variable in vector form as y where $y' = (y_1, y_2, y_3, ..., y_n)$. We denote the values predicted by the model, or fitted values, as $\hat{y}$ where $\hat{y}' = (\hat{y}_1, \hat{y}_2, \hat{y}_3, ..., \hat{y}_n)$. We will conclude that the model fits if (1) summary measures of the distance between y and $\hat{y}$ are small and (2) the contribution of each pair $(y_i, \hat{y}_i)$, $i = 1, 2, 3, ..., n$ to these summary measures is unsystematic and is small relative to the error structure of the model. Thus, a complete assesssment of the fitted model will involve both the calculation of summary measures of the distance between y and $\hat{y}$, and a thorough examination of the individual components of these measures.

The development of methods for assessment of goodness-of-fit will follow what we feel are the logical steps upon completion of the model building stage. The components of the proposed approach are (1) computation and evaluation of overall measures of fit, (2) examination of the individual components of the summary statistics, often graphically, and (3) examination of other measures of the difference or distance between the components of y and $\hat{y}$.

5.2 Summary Measures of Goodness-of-Fit

We begin with the summary measures of goodness-of-fit, as they are routinely provided as output with any fitted model and give an overall indication of the fit of the model. Because these are summary statistics, they may not be very specific about the individual components. A small value for one of these statistics does not rule out the possibility of some substantial and thus interesting deviation from fit for a few subjects. On the other hand, a large value for one of these statistics is a clear indication of a substantial problem with the model.

Before discussing specific goodness-of-fit statistics, we must examine the effect the fitted model has on the available information as characterized by the degrees of freedom. We will use the term **covariate pattern** to describe a single set of values for the covariates in a model. At the model development stage we assumed that the values of the independent variables were such that each subject was unique in their configuration of the covariates. That is, the number of covariate patterns was equal to n. For example, if age, race, sex and weight were our variables, then the combination of these may well result in a unique set of values for each subject. Once a final model is obtained there may be relatively few variables in the model, and the number of covariate patterns may be less than n. For example, if the final model contains only race and sex, each coded at two levels, then there are only four possible covariate patterns.

At the model development stage it is not necessary to be concerned about the number of covariate patterns, as they are determined by the richest possible model and are thus common to all models. The degrees of freedom for tests are based on the difference in the number of variables in competing models, not on the number of covariate patterns. However, the number of covariate patterns may be an issue when the fit of a model is assessed.

Goodness-of-fit is assessed over the constellation of fitted values determined by the covariates in the model, not the total collection of covariates. For instance, suppose that our fitted model contains p independent variables, $\mathbf{x}' = (x_1, x_2, x_3, ..., x_p)$, and let J denote the number of distinct values of $\mathbf{x}$ observed. If some subjects have the same value of $\mathbf{x}$ then $J < n$. We will denote the number of subjects with $\mathbf{x} = \mathbf{x}_j$ by m_j, $j = 1, 2, 3, ..., J$. It follows that $\Sigma m_j = n$. Let y_j denote the number of positive responses, $y = 1$, among the m_j subjects with $\mathbf{x} = \mathbf{x}_j$. It follows that $\Sigma y_j = n_1$, the total number of subjects with $y = 1$. The distribution of the goodness-of-fit statistics is obtained by letting n become large. If the number of covariate patterns also increases with n then each value of m_j will tend to be small. Distributional results obtained under the condition that only n becomes large are said to be based on n-**asymptotics**. If we fix $J < n$ and let n become large then each value of m_j will tend to become large. Distributional results based on each m_j becoming large are said to be based on m-**asymptotics**. The difference and need to distinguish between these two asymptotics will become clearer when summary statistics are discussed in the next section.

Initially, we will assume that $J \approx n$ as this is the case most frequently occurring and the one that provides the greatest difficulty in developing the distribution of goodness-of-fit statistics. We expect to have $J \approx n$ whenever we have some continuous covariates in the model.

5.2.1 Pearson Chi-Square and Deviance

In linear regression, summary measures of the distance between observed and fitted values, as well as diagnostics for casewise effect on the fit, are functions of a residual defined as the difference betweeen the observed and fitted value. That is, the residual $= (y - \hat{y})$. In logistic regression there are several possible measures of difference between the observed and fitted values. To emphasize the fact that the fitted values in logistic regression are calculated for each covariate pattern and depend on the estimated probability for that covariate pattern, we will denote the fitted value, $\hat{y}_j$, as

$$m_j\hat{\pi}_j = m_j(\exp[\hat{g}(\mathbf{x}_j)]/\{1 + \exp[\hat{g}(\mathbf{x}_j)]\})$$

where $\hat{g}(\mathbf{x}_j)$ is the estimated logit.

We begin by considering two measures of the difference between the observed and the fitted values: the Pearson residual and deviance residual. For a particular covariate pattern the Pearson residual is defined as follows:

$$r(y_j,\hat{\pi}_j) = \frac{(y_j - m_j\hat{\pi}_j)}{\sqrt{m_j\hat{\pi}_j(1 - \hat{\pi}_j)}} \tag{5.1}$$

The summary statistic based on these residuals is the Pearson chi-square statistic

$$X^2 = \sum_{j=1}^{J} r(y_j,\hat{\pi}_j)^2 \tag{5.2}$$

The deviance residual is defined as follows:

$$d(y_j,\hat{\pi}_j) = \pm\left\{ 2\left[y_j \ln\left(\frac{y_j}{m_j\hat{\pi}_j}\right) + (m_j - y_j) \ln\left(\frac{(m_j - y_j)}{m_j(1 - \hat{\pi}_j)}\right) \right] \right\}^{1/2} \tag{5.3}$$

where the sign is the same as the sign of $(y_j - m_j\hat{\pi}_j)$. For covariate patterns with $m_j = 1$ and y_j the deviance residual is

$$d(y_j,\hat{\pi}_j) = -\sqrt{2|\ln(1 - \hat{\pi}_j)|}$$

and the deviance residual, when $m_j = 1$ and $y_j = 1$, is

$$d(y_j,\hat{\pi}_j) = \sqrt{2|\ln(\hat{\pi}_j)|}$$

The summary statistic based on the deviance residuals is the deviance

$$D = \sum_{j=1}^{J} d(y_j,\hat{\pi}_j)^2 \tag{5.4}$$

This is the same quantity shown in equation (1.10).

The distribution of the statistics X^2 and D under the assumption that the fitted model is correct in all aspects is supposed to be chi-square with degrees of freedom equal to $J - (p + 1)$. For the deviance this statement follows from the fact that D is the likelihood ratio test statistic of a saturated model with J parameters versus the fitted model with $p + 1$ parameters. Similar theory provides the null distribution of X^2. The problem is that when $J \approx n$, the distribution is obtained under n-asymptotics, and hence the number of parameters is increasing at the same rate as the sample size. Thus, p-values calculated for these two statistics when $J \approx n$, using the $\chi^2(J - p - 1)$ distribution, are incorrect. McCullagh and Nelder (1983) examine the expected value of X^2 and D when the parameters of the model have been estimated. They show that, in this situation, the expected value is less than the stated degrees of freedom, $J - (p + 1)$. They provide a correction factor that may be used when $m_j \hat{\pi}_j$ and $m_j(1 - \hat{\pi}_j)$ both exceed one for each j. Our experience, albeit limited, with application of their correction factor when $J \approx n$ suggests that it provides too great a reduction in the expected value. For practical purposes this suggests that when $J \approx n$ use of $J - (p + 1)$ provides a reasonable estimate of the expected value of X^2 and D, when the model fit is the correct one.

As an example, consider the final fitted model for the low birth weight data given in Table 4.9. The 10 variables included in the model are: AGE, RACE (categorical at three levels), smoking status (SMOKE), hypertension (HT), uterine irritabilty (UI), previous preterm labor (PTD), low weight at the last menstrual period (LWD), the interaction of age and low weight (AGE × LWD), and the interaction of smoking and low weight (SMOKE × LWD). There are $J = 128$ covariate patterns for this data set. The values of the two statistics are $X^2 = 137.4$ and $D = 150.6$, with degrees of freedom equal to $128 - 11 = 117$.

One way to avoid the above noted difficulties with the distributions of X^2 and D when $J \approx n$ is to group the data in such a way that m-asymptotics can be used. To understand the rationale behind the various grouping strategies that have been proposed, it is helpful to think of X^2 as the Pearson and D as the log-likelihood chi-square statistics which result from a $2 \times J$ table. The rows of the table correspond to the two values of the outcome variable, $y = 1, 0$. The J columns correspond to the J possible covariate patterns. The estimate of the expected value under the hypothesis that the logistic model in question is the correct model for the cell corresponding to the $y = 1$ row and jth column is $m_j \hat{\pi}_j$. It follows that the estimate of the expected value for the cell corresponding to the $y = 0$ row and jth column is $m_j(1 - \hat{\pi}_j)$. The statistcs X^2 and D are calculated in the usual manner from this table.

Thinking of the statistics as arising from the $2 \times J$ table gives some intuitive insight as to why we cannot expect them to follow the $\chi^2(J - p - 1)$ distribution. When chi-square tests are computed from a contingency table the p-values are correct under the hypothesis when the estimated expected values are "large" in each cell. This conditon holds under m-asymptotics. Although this is an oversimplification of the situation, it is essentially correct. In the above described $2 \times J$ table the expected values will always be quite small since the number of columns increases as n increases. To avoid this problem we may collapse the columns into a fixed number of groups, g, and then calculate observed and expected frequencies. By fixing the number of columns, the estimated expected frequencies will become large as n becomes large. Thus, m-asymptotics hold. The theory required to derive the distribution of the statistics is not quite so straightforward but the intuitive appeal of thinking in this manner is most helpful. The relevant distribution theory presented in a series of papers by Moore (1971), and Moore and Spruill (1975), considers what happens to chi-square goodness-of-fit tests when the boundaries forming the cells are functions of random variables.

5.2.2 The Hosmer–Lemeshow Tests

Hosmer and Lemeshow (1980) and Lemeshow and Hosmer (1982) proposed grouping based on the values of the estimated probabilities. Suppose for sake of discussion, that $J = n$. In this case we think of the n columns as corresponding to the n values of the estimated probabilities, with the first column corresponding to the smallest value, and the nth column to the largest value. Two grouping strategies were proposed as follows: (1) collapse the table based on percentiles of the estimated probabilities and (2) collapse the table based on fixed values of the estimated probability.

With the first method, use of $g = 10$ groups results in the first group containing the $n_1' = n/10$ subjects having the smallest estimated probabilities, and the last group containing the $n_{10}' = n/10$ subjects having the largest estimated probabilities. With the second method, use of $g = 10$ groups results in cutpoints defined at the values $k/10$, $k = 1, 2, ..., 9$, and the groups contain all subjects with estimated probabilities between adjacent cutpoints. For example, the first group contains all subjects whose estimated probability is less than or equal to 0.1, while the tenth group contains those subjects whose estimated probability is greater then 0.9. For the $y = 1$ row, estimates of the expected values are obtained by summing the estimated probabilites over all subjects in a group. For the $y = 0$ row, the estimated expected value is obtained by summing, over all subjects in the group, one minus the estimated probability. For either grouping strategy, the Hosmer–Lemeshow goodness-of-fit statistic, $\hat{C}$, is obtained by calculating the Pearson chi-square statistic from the $2 \times g$ table of observed and estimated expected frequencies. A formula defining the calculation of $\hat{C}$ is as follows:

$$\hat{C} = \sum_{k=1}^{g} \frac{(o_k - n_k' \bar{\pi}_k)^2}{n_k' \bar{\pi}_k (1 - \bar{\pi}_k)} \tag{5.5}$$

where n_k' is the number of covariate patterns in the kth group,

$$o_k = \sum_{j=1}^{n_k'} y_j$$

the number of responses among the n_k' covariate patterns, and

$$\bar{\pi}_k = \sum_{j=1}^{n_k'} m_j \hat{\pi}_j / n_k'$$

the average estimated probability.

Using an extensive set of simulations, Hosmer and Lemeshow (1980) demonstrated that, when $J = n$ and the fitted logistic regression model is the correct model, the distribution of the statistic $\hat{C}$ is well approximated by the chi-square distribution with $g - 2$ degrees of freedom, $\chi^2(g - 2)$. While not specifically examined it is likely that $\chi^2(g - 2)$ will also approximate the distribution when $J \approx n$.

An alternative to the denominator shown in equation (5.5) is obtained if we consider o_k to be the sum of independent nonidentically distributed random variables. This suggests that we should standardize the squared difference between the observed and estimated expected frequency by

$$\sum_{j=1}^{n_k'} \hat{\pi}_j (1 - \hat{\pi}_j)$$

It is easy to show that

$$\sum_{j=1}^{n_k'} \hat{\pi}_j(1 - \hat{\pi}_j) = n_k' \bar{\pi}_k(1 - \bar{\pi}_k) - \sum_{j=1}^{n_k'} (\hat{\pi}_j - \bar{\pi}_k)^2$$

We have found that the use of the expression

$$\sum_{j=1}^{n_k'} \hat{\pi}_j(1 - \hat{\pi}_j)$$

usually results in a trivial increase in the value of the test statistic. Thus, in practice we calculate $\hat{C}$ using equation (5.5).

Additional research by Hosmer, Lemeshow, and Klar (1988) has shown that the grouping method based on percentiles of the estimated probabilities is preferable to the one based on fixed cutpoints in the sense of better adherence to the $\chi^2(g - 2)$ distribution, especially when many of the estimated probabilities are small (i.e., less than 0.2). Thus, unless stated otherwise, $\hat{C}$ will be based on the percentile-type of grouping, and usually with $g = 10$ groups. These groups are often referred to as the "deciles of risk." This term comes from health sciences research where the outcome $y = 1$ often represents the occurrence of some disease. This statistic may be obtained without much difficulty from any package with reasonable data-handling capabilities which gives access to the estimated probabilities. One program, BMDPLR, provides this statistic as part of the output for fitted logistic regression models.

The results of applying the decile of risk grouping strategy to the estimated probabilities computed from the model for low birth weight given in Table 4.9 are given in Table 5.1. For example, the observed frequency in the low weight ($y = 1$) group for the fifth decile of risk is 6. This value is obtained from the sum of the observed outcomes for the 20 subjects in this group. In a similar fashion the corresponding estimated expected frequency for this decile is 5.02, which is the sum of the 20 estimated probabilities for these subjects. The observed frequency for the normal weight ($y = 0$) group is $20 - 6 = 14$, and the estimated expected frequency is $20 - 5.02 = 14.98$.

Table 5.1 Observed (Obs) and Estimated Expected (Exp) Frequencies Within Each Decile of Risk for Each Outcome, Low Weight, and Normal Weight Using the Fitted Logistic Regression Model in Table 4.9.

Weight		\multicolumn{10}{c}{Decile of Risk}										Total
		1	2	3	4	5	6	7	8	9	10	
Low	Obs	0	1	4	2	6	6	6	10	9	15	59
	Exp	0.9	1.6	2.3	3.7	5.0	5.6	6.8	8.6	10.5	14.1	59
Normal	Obs	18	19	14	18	14	12	12	9	10	4	130
	Exp	17.2	18.4	15.8	16.4	15.0	12.4	11.2	10.4	8.5	4.9	130
	Total	18	20	18	20	20	18	18	19	19	19	189

The value of the Hosmer–Lemeshow goodness-of-fit statistic computed from the frequencies in Table 5.1 is $\hat{C} = 5.23$, and the corresponding p-value computed from the chi-square distribution with 8 degrees of freedom is 0.73. This indicates that the model seems to fit quite well. A comparison of the observed and expected frequencies in each of the 20 cells in Table 5.1 shows that the model fits within each decile of risk.

Because the distribution of $\hat{C}$ depends on m-asymptotics the appropriateness of the p-value will depend on the estimated expected frequencies being large enough to employ this theory. Examining the estimated expected frequencies we see that only one is less than 1 and five are less than 5. Whether this is of concern depends on how one regards the magnitude of estimated expected frequencies in contingency tables. Our point of view is more liberal than those who maintain that with tables of this size (about 20 cells), all expected frequencies must be greater than 5. In this case, we feel that there is reason to believe that the calculation of the p-value is accurate enough to support the hypothesis that the model fits. If one is concerned about the magnitude of the expected frequencies, selected adjacent columns of the table may be combined to increase the size of the expected frequencies while, at the same time, reducing the number of degress of freedom.

A few additional comments about the calculation of $\hat{C}$ are needed. When the number of covariate patterns is less than n, we have the possibility that one or more of the empirical deciles will occur at a pattern with $m_j > 1$. If this happens, the value of $\hat{C}$ will depend on how these ties are assigned. In Table

5.1, ties are assigned to deciles in such as way as to make the column totals as close to $n/10$ as possible. In contrast to this strategy BMDPLR always assigns ties to the same decile. This strategy could result in fewer than 10 groups and, consequently, fewer degrees of freedom. For example, the value of $\hat{C}$ obtained from BMDPLR for the model in Table 4.9 is 4.32, compared to the value 5.23 obtained from Table 5.1, each with 8 degrees of freedom. The merits of the two grouping strategies has not been studied in great enough detail to be able to recommend one method over the other.

If too few groups are used to calculate $\hat{C}$, then we run the risk that we will not have the sensitivity to be able to distinguish observed from expected frequencies. It has been our experience that when $\hat{C}$ is calculated from fewer than 6 groups it will almost always indicate that the model fits.

The advantage of a summary goodness-of-fit statistic like $\hat{C}$ is that it provides a single, easily interpretable, value which can be used to assess fit. The great disadvantage is that in the process of reducing the data through grouping we may miss an important deviation from fit due to a small number of individual data points. Hence we advocate that, before finally accepting that a model fits, an analysis of the individual residuals and relevant diagnostic statistics be performed. These methods will be presented in the next section.

Our experience is that a table such as the one presented in Table 5.1 contains valuable descriptive information for assessing the adequacy of the fitted model over the deciles of risk. For example, comparison of observed to expected frequencies within each cell may indicate regions where the model does not perform satisfactorily.

Other grouping strategies have been proposed which lead to statistics similar to $\hat{C}$. Tsiatis (1980) suggested a goodness-of-fit statistic based on an explicit partition of the covariate space into g regions. A categorical variable with g levels is introduced into the model corresponding to the g groups. The goodness-of-fit test is the Score test of the coefficients for the new grouping variable. Tsiatis showed that the Score test for this variable is based on a comparison of the observed frequency to estimated expected frequency within each of the g groups. The test has $g - 1$ degrees of freedom. This test can be easily carried out in the EGRET, GLIM, or SAS packages. When it is difficult or unclear how to partition the covariate space into meaningful groups, then an alternative to explicit partitioning is to use deciles of risk. Application of the Score test to assess the fit of the model in Table 4.9 using the deciles of risk shown in Table 5.1 yields a value of 9.48 which, with 9 degrees of freedom, gives a p-value of 0.39. Hence, this test also supports the fit of the model. One disadvantage of using the Score test is that actual values of the observed and estimated expected frequencies need not be obtained. These quantities may be useful, when there is evidence of lack of fit, in indicating those deciles where it is occurring.

5.2.3 Other Summary Statistics

Other summary statistics have been proposed for goodness-of-fit which are not based directly on the distance between the observed and fitted values. These include a Score test proposed by Brown (1982), and various measures of classification accuracy.

The statistic proposed by Brown is based on embedding the logistic regression model into a larger family of generalized response models first suggested by Prentice (1976). This larger class has two additional parameters which, when equal to 1, yield the logistic regression model. Brown (1982) derived the Score test for these additional two parameters. The details for computation of the test may be found in Brown's paper. The test is part of the standard output of program BMDPLR. A large value for this test would indicate that the two additional parameters in the extended model may be different from one and that the extended model fits better than the logistic model. A small value would indicate that the two parameters in the extended model may not be different from one and its fit is not better than the logistic model. The Brown test provides a relative comparison of two fitted models, not an absolute comparison of the fitted values of one model to the observed or fitted values of the saturated model. In our opinion only a comparison of the latter type provides a correct assessment of goodness-of-fit.

The value of the Brown test for the model given in Table 4.9 is 2.03 which, with 2 degrees of freedom, yields a p-value of 0.36. This indicates that the two additional parameters of the expanded model may not be different from one.

An intuitively appealing way to summarize the results of a fitted logistic model is via a classification table. This table is the result of cross-classifying the outcome variable, y, with a dichotomous variable whose values are derived from the estimated logistic probabilities.

To obtain the derived dichotomous variable we must define a cutpoint, c, and compare each estimated probability to c. If the estimated probability exceeds c then we let the derived variable be equal to 1; otherwise it is equal to zero. The most commonly used value for c is 0.5. The appeal of this type of approach to model assessment comes from the close relationship of logistic regression to discriminant analysis when the distribution of the covariates is multivariate normal within the two outcome groups. However, it is not limited to this model; see, for example, Efron (1975).

In this approach, estimated probabilities are used to predict group membership. Presumably, if the model predicts group membership accurately according to some criterion, then this is thought to provide evidence that the model fits. Unfortunately, this may or may not be the case. For example, it is easy to construct a situation where the logistic regression model is in fact the correct model and thus will fit, but classification will be poor. Suppose that $P(Y = 1) = \theta_1$ and that $X \sim N(0,1)$ in the group with $Y = 0$ and and $X \sim N(\mu,1)$ in the group with $Y = 1$. In this model the slope coefficient for the logistic regression model is $\beta_1 = \mu$ and the intercept is

$$\beta_0 = \ln\left[\frac{\theta_1}{(1 - \theta_1)}\right] - \frac{\mu^2}{2}$$

The probability of misclassification, PMC, is

$$\text{PMC} = \theta_1 \Phi\left\{\frac{1}{\beta_1} \ln\left[\frac{(1 - \theta_1)}{\theta_1}\right] - \frac{\beta_1}{2}\right\}$$

$$+ (1 - \theta_1)\Phi\left\{\frac{1}{\beta_1} \ln\left[\frac{\theta_1}{(1 - \theta_1)}\right] - \frac{\beta_1}{2}\right\}$$

where Φ is the cumulative distribution function of the N(0,1) distribution. Thus, the expected error rate is a function of the magnitude of the slope, not necessarily of the fit of the model. Accurate or inaccurate classification does not address our criteria for goodness-of-fit; that the distances between observed and expected values be unsystematic, and within the variation of the model. However, the classification table may be a useful adjunct to other measures based more directly on residuals.

The results of classifying the observations of low birth weight using the fitted model given in Table 4.9 are presented in Table 5.2.

Table 5.2 Classification Table Based on the Logistic Regression Model in Table 4.9.

Observed	Classified		
	Low Weight	Normal Weight	Total
Low Weight	24	35	59
Normal Weight	14	116	130
Total	38	151	189

The classification table shown in Table 5.2 is fairly typical of those seen in many logistic regression applications. The overall rate of correct classification is estimated as 74.1 = 100[(24 + 116)/189]%, with 89.2% of the normal weight group, and only 40.7% of the low weight group being correctly classified. Classification is sensitive to the relative sizes of the two component groups and will always favor classification into the larger group, a fact that is also independent of the fit of the model. This may be seen by considering the expression for PMC as a function of θ_1. The disadvantage of using PMC as a criterion is that it reduces a probablistic model where outcome is measured on a continuum, to a dichotomous model where predicted outcome is binary. For practical purposes there is little difference between the values of $\hat{\pi} = 0.48$ and $\hat{\pi} = 0.52$, yet use of a 0.5 cutpoint would establish these two individuals as markedly different. In summary, the classification table is most appropriate when classification is a stated goal of the analysis; otherwise it should only supplement more rigorous methods of assessment of fit.

For sake of completeness we present an R^2-type measure which has been suggested for use with logistic regression. For reasons we will explain later, we do not advocate the use of this statistic for assessing goodness-of-fit.

In linear regression, R^2 is the ratio of the regression sum-of-squares to the total sum-of-squares. It is convenient to think of the regression sum-of-squares as being the difference between the total and residual sum-of-squares. Under the assumption that the errors in the linear regression model are normal, the residual sum-of-squares for a fitted model is proportional to the log-likelihood for the model.

Let L_0 and L_p denote the log-likelihoods for models containing only the intercept, and the model containing the intercept plus the p covariates, respectively. Hence, in linear regression

$$R^2 \approx 100(L_0 - L_p)/L_0 = 100(1 - L_p/L_0)$$

This form of the statistic is proposed for use in logistic regression. The value of this statistic, when computed using log-likelihoods from logistic regression, will be denoted as R_L^2. (The subscript "L" has been added to emphasize that we are calculating a quantity specific to logistic regression.) The maximum value for R_L^2 is obtained when we fit the saturated model. In this case, we see that R_L^2 will be equal to 100% when $J = n$; otherwise the maximum will be less than 100%.

The value of the log-likelihood from the saturated model, L_s, may be easily obtained from the deviance for the model with p covariates and its log-likelihood as $L_s = L_p + 0.5D$. Hence, it would seem prudent to calculate L_s whenever $J < n$ and to make comparisons relative to it. A modification of R_L^2 which incorporates L_s is

$$R_L^2 = 100(L_0 - L_p)/(L_0 - L_s) \tag{5.6}$$

which is the form of this statistic we will use.

For the low birth weight data the value of the log-likelihood for the intercept only model is $L_0 = -117.34$ and for the model in Table 4.9 is $L_p =$

-96.01. The deviance is $D = 150.6$. Thus, the value of the log likelihood for the saturated model is

$$L_s = -96.01 + 0.5 \times 150.6 = -20.71$$

and

$$R_L^2 = 100\{[-117.34 - (-96.01)]/[-117.34 - (-20.71)]\} = 22.1$$

Interpretation of R_L^2 requires that we be able to interpret the quantities in equation (5.6). For a fixed set of possible covariates, more than p, the quantity in the denominator of equation (5.6) is constant and the numerator is one-half the likelihood ratio test for the significance of the slope coefficients for the p covariates in the fitted model. Thus, the quantity R_L^2 is nothing more than an expression of the likelihood ratio test and, as such, is not a measure of goodness-of-fit. This likelihood ratio test compares fitted values under two models rather than comparing observed values to those fitted under one model.

5.3 Logistic Regression Diagnostics

The summary statistics described in the previous section, based on the Pearson and/or deviance residuals, provide a single number which summarizes the agreement of observed and fitted values. The advantage (as well as the disadvantage) of these statistics is that a single number is used to summarize considerable information. Therefore, before "accepting" that the model fits, it is crucial that other measures be examined to see if fit is supported over the entire set of covariate patterns. This is accomplished through a series of specialized measures falling under the general heading of regression diagnostics. We assume that the reader has had some experience with diagnostics for linear regression. For a brief introduction to linear regression diagnostics see Kleinbaum, Kupper, and Muller (1987). A more detailed presentation may be found in Cook and Weisberg (1982) and Belsley, Kuh, and Welsch (1980). Pregibon (1981) provided the theoretical work which extended linear regression diagnostics to logistic regression. Since that key paper work has focused on refining the use of logistic regression diagnostics in assessing goodness-of-fit. The relevant literature will be cited as we come to specific contributions and methods. We begin by briefly describing logistic regression diagnostics. In this development we assume that the fitted model contains p covariates and that they form J covariate patterns indexed by $j = 1, 2, ..., J$. The derivation of the diagnostics is at a higher mathematical level than most of the material in this text. It is not necessary to understand the mathematical development to effectively apply the diagnostics in practice. Thus, the less sophisticated mathematical reader may wish to skip to the discussion of Figure 5.1 where the discussion of the calculation and use of the diagnostics begins.

The key quantities for logistic regression diagnostics, as in linear regression, are the components of the "residual sum-of-squares." In linear regression a key assumption is that the error variance does not depend on the conditional mean, $E(Y_j | \mathbf{x}_j)$. However, in logistic regression we have binomial errors and, as a result, the error variance is a function of the conditional mean: $\text{var}(Y_j | \mathbf{x}_j) = m_j E(Y_j | \mathbf{x}_j) \times [1 - E(Y_j | \mathbf{x}_j)] = m_j \pi(\mathbf{x}_j)[1 - \pi(\mathbf{x}_j)]$. Thus, we begin with residuals as defined in equation (5.1) and (5.3) which have been "divided" by estimates of their standard errors; this may not be entirely obvious in the case of the deviance residual. Let r_j and d_j denote the values of the expressions given in equation (5.1) and (5.3), respectively, for covariate pattern $\mathbf{x}_j$. Since each residual has been divided by an approximate estimate of its standard error, we expect that if the logistic regression model is correct these quantities will have a mean approximately equal to zero and a variance approximately equal to 1. We will discuss their distribution shortly.

Besides the residuals for each covariate pattern other quantities central to the formation and interpretation of linear regression diagnostics are the "hat" matrix and the leverage values derived from it. In linear regression the hat matrix is the matrix that provides the fitted values as the projection of the outcome variable into the covariate space. Let $\mathbf{X}$ denote the $J \times (p + 1)$ matrix containing the values for all J covariate patterns formed from the observed values of the p covariates, with the first column being one to reflect the presence of an intercept in the model. The matrix $\mathbf{X}$ is often called the design matrix. In linear regression the hat matrix is $\mathbf{H} = \mathbf{X}(\mathbf{X}'\mathbf{X})^{-1}\mathbf{X}'$; for example, $\hat{\mathbf{y}} = \mathbf{H}\mathbf{y}$. The linear regression residuals, $(\mathbf{y} - \hat{\mathbf{y}})$, expressed in terms of the hat matrix are $(\mathbf{I} - \mathbf{H})\mathbf{y}$ where $\mathbf{I}$ is the $J \times J$ identity matrix. Using weighted least squares linear regression as a model, Pregibon (1981) derived a linear approximation to the fitted values which yields a hat matrix for logistic regression. This matrix is

$$\mathbf{H} = \mathbf{V}^{1/2}\mathbf{X}(\mathbf{X}'\mathbf{V}\mathbf{X})^{-1}\mathbf{X}'\mathbf{V}^{1/2} \tag{5.7}$$

where $\mathbf{V}$ is a $J \times J$ diagonal matrix with general element $v_j = m_j\hat{\pi}(\mathbf{x}_j)[1 - \hat{\pi}(\mathbf{x}_j)]$.

In linear regression the diagonal elements of the hat matrix are called the leverage values and are proportional to the distance of $\mathbf{x}_j$ to the mean of the data. This concept of distance to the mean is important in linear regression, as points that are far from the mean may have considerable influence on the values of the estimated parameters. The extension of the concept of leverage to logistic regression requires additional discussion and clarification.

Let the quantity h_j denote the jth diagonal element of the matrix $\mathbf{H}$ defined in equation (5.7). It may be shown that

$$h_j = m_j\hat{\pi}(\mathbf{x}_j)[1 - \hat{\pi}(\mathbf{x}_j)](1,\mathbf{x}_j')(\mathbf{X}'\mathbf{V}\mathbf{X})^{-1}(1,\mathbf{x}_j')' = v_j \times b_j \tag{5.8}$$

where

$$b_j = (1,\mathbf{x}_j')(\mathbf{X}'\mathbf{V}\mathbf{X})^{-1}(1,\mathbf{x}_j')'$$

and, as is the case in linear regression, $\Sigma h_j = (p + 1)$ the number of parameters in the model. In linear regression the dimension of the hat matrix is usually taken to be $n \times n$, and thus ignores any common covariate patterns in the data. With this formulation any diagonal element in the hat matrix has an upper bound of $1/k$, where k is the number of subjects with the same covariate pattern. If we formulate the hat matrix for logistic regression as an $n \times n$ matrix then each diagonal element is bounded from above by $1/m_j$ where m_j represents the total number of subjects with the same covariate pattern. When the hat matrix is based upon data grouped by covariate patterns, the upper bound for any diagonal element is 1. This distinction is important to keep in mind as diagnostic information is computed differently in various programs. For example, unless the data have been grouped, the hat matrix elements will be based on an $n \times n$ matrix in GLIM, whereas diagnostics in BMDPLR are always based on the number of covariate patterns. Thus, with BMDPLR, if consideration is restricted to those variables in the final model, diagnostics based on the hat matrix will use the correct $J \times J$ matrix. To avoid confusion and possible erroneous results, we recommend that when the number of covariate patterns, J, is much less than n or some values of m_j are larger than 5 the outcome variable be aggregated by covariate pattern prior to computation and analysis of diagnostic statistics.

In the final model for the low birth weight data shown in Table 4.9 we have $J = 128$ and $n = 189$. In this situation we definitely should aggregate the data before computing the diagnostic statistics. If, on the other hand, we had a model with $J = 175$ and $n = 189$, we might not go to the trouble to aggregate the data. When the number of covariate patterns is much less than n there is the risk that we may fail to identify influential and/or poorly fit covariate patterns.

Consider a covariate pattern with m_j subjects, $y_j = 0$ and estimated logistic probability $\hat{\pi}_j$. The Pearson residual defined in equation (5.1) computed for each individual subject with this covariate pattern is

$$r_i = \frac{(0 - \hat{\pi}_j)}{\sqrt{\hat{\pi}_j(1 - \hat{\pi}_j)}}$$

$$= -\sqrt{\frac{\hat{\pi}_j}{(1 - \hat{\pi}_j)}}$$

while the Pearson residual based on all subjects with this covariate pattern is

$$r_j = \frac{(0 - m_j\hat{\pi}_j)}{\sqrt{m_j\hat{\pi}_j(1 - \hat{\pi}_j)}}$$

$$= -\sqrt{m_j}\sqrt{\frac{\hat{\pi}_j}{(1 - \hat{\pi}_j)}}$$

which increases negatively as m_j increases. If $m_j = 1$ and $\hat{\pi}_j = 0.5$, then $r_j = -1$ which is not a large residual. On the other hand, if there were $m_j = 16$ subjects with this covariate pattern, then $r_j = -4.0$, which is quite large.

The steps necessary to form an aggregated data set will depend on the software package. It is easy to carry out in some and impossible in others. In these latter packages we will have to use a second package to create the aggregated data set and then return to our logistic regression package with this new data set. The essential steps in any package are: (1) Define as aggregation variables the main effects in the model. This defines the covariate patterns. (2) Calculate the sum of the outcome variable, and number of terms in the sum over the aggregation variables. This produces y_j and m_j for each covariate pattern. (3) Output a new data set containing the values of the aggregation variables, covariate patterns, and the two calculated variables, y_j and m_j.

A second point which must be kept in mind when interpreting the magnitude of h_j in equation (5.8) is the effect that v_j has on this value. Pregibon (1981) notes that the fit determines the estimated coefficients and, since the estimated coefficients determine the estimated probabilities, points with large values of h_j are extreme in the covariate space and thus lie far from the mean. This point was refuted by Lesaffre (1986), page 117, where he shows that the term v_j in the expression for h_j cannot be ignored. The following example will demonstrate that, up to a point, both Pregibon and Lesaffre are correct.

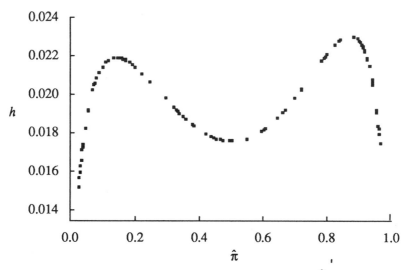

Figure 5.1 Plot of Leverage (h) Versus ($\hat{\pi}$).

Figure 5.1 presents a plot of the leverage values versus the estimated probabilities for a sample of 100 observations from a logistic model with $g(x) = 0.583x$ and $x \sim U(-6,6)$ Recall that the notation $U(a,b)$ describes a variable following the uniform distribution on the interval (a,b).

We see that the leverage increases as the estimated probability gets further from 0.5 (x gets further from its mean, nominally zero) until the estimated probabilities become less than 0.1 or greater than 0.9. At that point the leverage decreases and rapidly approaches zero. This example shows that the points most extreme in the covariate space may have the smallest leverage. This is the exact opposite of the situation in linear regression, where the leverage is a monotonic increasing function of the distance of a covariate pattern to the mean. The practical consequence of this is that to correctly interpret a particular value of the leverage in logistic regression, we need to know whether or not the estimated probability is small (<0.1) or large (>0.9). If the estimated probability lies between 0.1 and 0.9 then the leverage will give a value that may be thought of as distance. When the estimated probability lies outside the interval 0.1 to 0.9, then the value of the leverage may not measure distance in the sense that further from the mean implies a larger value.

A quantity that does increase with the distance from the mean is $b_j = (1,x_j')(\mathbf{X'VX})^{-1}(1,x_j')'$. Thus, if we are only interested in distance then we should focus on b_j. A plot of the b_j versus the estimated probability for the example is shown in Figure 5.2.

In this figure we see that b_j provides a measure of distance in the covariate space and, as a result, is more like the leverage values in linear regression. However, since the most useful diagnostic statistics for logistic regression are functions of h_j, b_j will not be discussed further here.

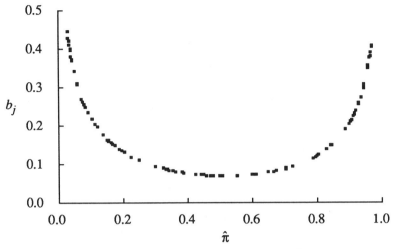

Figure 5.2 Plot of b_j Versus ($\hat{\pi}$).

If we consider the residual for the jth covariate pattern as $[y_j - m_j\hat{\pi}(\mathbf{x}_j)] \approx (1 - h_j)y_j$, then the variance of the residual is

$$m_j\hat{\pi}(\mathbf{x}_j)[1 - \hat{\pi}(\mathbf{x}_j)](1 - h_j)$$

which suggests that the Pearson residuals will not have variance equal to 1 unless they are further standardized. Recall that we denote by r_j the Pearson residual given in equation (5.1). The standardized Pearson residual for covariate pattern $\mathbf{x}_j$ is

$$r_{sj} = r_j / \sqrt{1 - h_j} \tag{5.9}$$

Another useful diagnostic statistic is one that examines the effect that deleting all subjects with a particular covariate pattern has on the value of the estimated coefficients and the overall summary measures of fit, X^2 and D. The change in the value of the estimated coefficients is analogous to the measure proposed by Cook (1977, 1979) for linear regression. It is obtained as the standardized difference between $\hat{\beta}$ and $\hat{\beta}_{(-j)}$, where these represent the respective maximum likelihood estimates computed using all J covariate patterns and excluding the m_j subjects with pattern $\mathbf{x}_j$, and thus standardizing via the

estimated covariance matrix of $\hat{\beta}$. Pregibon (1981) showed, to a linear
approximation, that this quantity for logistic regression is

$$\Delta\hat{\beta}_j = (\hat{\beta} - \hat{\beta}_{(-j)})'(\mathbf{X}'\mathbf{V}\mathbf{X})(\hat{\beta} - \hat{\beta}_{(-j)})$$

$$= \frac{r_j^2 h_j}{(1 - h_j)^2}$$

$$= \frac{r_{sj}^2 h_j}{(1 - h_j)} \tag{5.10}$$

Using similar linear approximations it can be shown that the decrease in the
value of the Pearson chi-square statistic due to deletion of the subjects with
covariate pattern $\mathbf{x}_j$ is

$$\Delta X_j^2 = \frac{r_j^2}{(1 - h_j)}$$

$$= r_{sj}^2 \tag{5.11}$$

A similar quantity may be obtained for the change in the deviance,

$$\Delta D_j = d_j^2 + \frac{r_j^2 h_j}{(1 - h_j)}$$

If we replace r_j^2 by d_j^2 this yields the approximation

$$\Delta D_j = \frac{d_j^2}{(1 - h_j)} \tag{5.12}$$

which is similar in form to the expression in equation (5.11).

These diagnostic statistics are conceptually quite appealing, as they allow
us to identify those covariate patterns that are poorly fit (large values of ΔX_j^2
and/or ΔD_j), and those that have a great deal of influence on the values of the
estimated parameters (large values of $\Delta\hat{\beta}_j$). After identifying these influential
patterns (subjects), we can begin to address the role they play in the analysis.

Before proceeding on to the use of the diagnostics in an example, we will
make a few summary comments on what we might expect their application to

tell us. Consider first the measure of fit ΔX_j^2. This measure will be smallest when y_j and $m_j \hat{\pi}(\mathbf{x}_j)$ are close. This is most likely to happen when $y_j = 0$ and $\hat{\pi}(\mathbf{x}_j) < 0.1$ or $y_j = m_j$ and $\hat{\pi}(\mathbf{x}_j) > 0.9$. Similarly ΔX_j^2 will be largest when y_j is furthest from $\hat{\pi}(\mathbf{x}_j)$. This is most likely to occur if we have a value of $y_j = 0$ and $\hat{\pi}(\mathbf{x}_j) > 0.9$, or with $y_j = m_j$ and $\hat{\pi}(\mathbf{x}_j) < 0.1$. These same covariate patterns are not likely to have a large $\Delta \hat{\beta}_j$ since, when $\hat{\pi}(\mathbf{x}_j) < 0.1$ or $\hat{\pi}(\mathbf{x}_j) > 0.9$, $\Delta \hat{\beta}_j \approx \Delta X_j^2 h_j$, and h_j is approaching zero. The influence diagnostic, $\Delta \hat{\beta}_j$, will be large when both ΔX_j^2 and h_j are at least moderate. This will most likely occur when $0.1 < \hat{\pi}(\mathbf{x}_j) < 0.3$, or $0.7 < \hat{\pi}(\mathbf{x}_j) < 0.9$. As we know from Figure 5.1, these are the intervals where h_j is largest. In the region where $0.3 < \hat{\pi}(\mathbf{x}_j) < 0.7$ the chances are not as great that either ΔX_j^2 or h_j will be large. Table 5.3 presents a summary of these observations. This table reports what might be expected, not what may actually happen in any particular example. Therefore, it should only be used as a guide to further understanding and interpretation of the diagnostic statistics.

Table 5.3   Likely Values of Each of the Diagnostic Statistics ΔX^2, $\Delta \hat{\beta}$, and h Within Each of Five Regions Defined by the Value of the Estimated Probability.

Diagnostic	Estimated Logistic Probability				
	0.0 – 0.1	0.1 – 0.3	0.3 – 0.7	0.7 – 0.9	0.9 – 1.0
ΔX^2	Large or Small	Moderate	Moderate to Small	Moderate	Large or Small
$\Delta \hat{\beta}$	Small	Large	Moderate	Large	Small
h	Small	Large	Moderate to Small	Large	Small

In linear regression essentially two approaches are used to interpret the value of the diagnostics and these are often used in conjunction with each other. The first is graphical. The second employs the distribution theory of the linear regression model to develop the distribution of the diagnostics under the assumption that the fitted model is correct. In the graphical approach, large values of diagnostics either appear as spikes or in the extreme corners of plots. A value of the diagnostic statistic for a point appearing to lie away from the balance of the points is judged to be extreme if it exceeds some percentile of the relevant distribution. This may sound a little too hypothesis-testing oriented;

but under the assumptions of linear regression with normal errors there is a known statistical distribution whose percentiles provide some guidance as to what constitutes a large value. Presumably, if the model is correct and fits then no values should be exceptionally large, and the plots should appear as expected under the distribution of the diagnostic.

In logistic regression we will have to rely primarily on visual assessment, as the distribution of the diagnostics under the hypothesis that the model fits is known only in certain limited settings. For instance, consider the Pearson residual, r_j. It is often stated that the distribution of this quantity will be approximately $N(0,1)$ when the model is correct. This statement is only true when m_j is sufficiently large to justify that the normal distribution provides an adequate approximation to the binomial distribution, a condition obtained under m-asymptotics. For example, if $m_j = 1$ then r_j will have only two possible values and can hardly be expected to be normally distributed. Jennings (1986b) has stated this point clearly and with all the necessary technical details. All of the diagnostics are evaluated by covariate pattern; hence any approximations to their distributions based on the normal distribution will, under binomial errors, depend on the number of subjects with that pattern. When a fitted model contains some continuous covariates that the number of covariate patterns, J, will be of the same order as n, and m-asymptotic results can not be relied upon. Thus, in practice, an assessment of "large" will of necessity be a judgment call based on experience and the particular set of data being analyzed. Using the $N(0,1)$, or equivalently, the $\chi^2(1)$ distribution for squared quantities may provide some guidance as to what large is. However, we urge that these percentiles be used with extreme caution.

We have defined seven diagnostic statistics which may be divided into three categories: (1) the basic building blocks, which are of interest in themselves, but also are used to form other diagnostics, (r_j, d_j, h_j); (2) derived measures of the effect of each covariate pattern on the fit of the model, $((r_{sj}, \Delta X_j^2, \Delta D_j)$; and (3) a derived measure of the effect of each covariate pattern on the value of the estimated parameters, $(\Delta \hat{\beta}_j)$. In practice, to use these diagnostic statistics we either need to have access to enough elements of the fitted model to compute the basic building blocks or to use a program that calculates them for us. The

GLIM package is an example of the former, while BMDPLR is an example of the latter.

A number of the logistic regression diagnostics may be obtained from any linear regression program that computes linear regression diagnostics. The reason for considering this is that while not all logistic regression packages provide the user with enough information to calculate the diagnostics, they may provide linear regression diagnostics. In addition, there are packages that have superb linear regression diagnostics, but no logistic regression capabilities, for example, Data Desk [Velleman and Velleman (1988)]. The results presented here follow directly from the least squares formulation presented in Section 4.4.

The pseudodependent variable, z, and weights, v, defined in Section 4.4 are based on the covariate patterns in our final model. Specifically, for each covariate pattern we compute

$$z_j = (1, \mathbf{x}_j') \hat{\boldsymbol{\beta}} + \frac{(y_j - m_j \hat{\pi}_j)}{m_j \hat{\pi}_j (1 - \hat{\pi}_j)}$$

and

$$v_j = m_j \hat{\pi}_j (1 - \hat{\pi}_j)$$

Let s denote the square root of the mean weighted residual sum of squares from the linear regression of z on the covariates in the final model, with case weight v. In this case the covariates, $\mathbf{x}$, contain all main effect as well as any interaction terms in the final model. It is easily shown that the leverage values from the least squares program are identical to those for logistic regression given in equation (5.8). That is,

$$h_j = \text{leverage from the weighted least squares fit}$$

Other diagnostics may be computed as follows:

$$t_j = \sqrt{v_j} \, (z_j - \hat{z}_j)$$

$$= \sqrt{v_j} \times (\text{residual from the weighted least squares fit})$$

$$r_{sj} = \sqrt{v_j} \, (z_j - \hat{z}_j) / \sqrt{(1 - h_j)}$$

$$= s \times \text{(studentized residual from the weighted least squares)}$$

$$\Delta X_j^2 = r_{sj}^2$$

$$= s^2 \times \text{(studentized residual from the weighted least squares)}^2$$

and

$$\Delta \hat{\beta}_j = (p + 1) \times s^2 \times \text{(Cook distance from the weighted least squares)}$$

Some linear regression packages do not have the option for a case weight variable. It is still possible to obtain the logistic regression diagnostics from these packages, but we must weight the data before running the least squares regression program. In this situation we calculate the square root of the weight,

$$w_j = \sqrt{m_j \hat{\pi}_j (1 - \hat{\pi}_j)}$$

$$= \sqrt{v_j}$$

and then multiply z_j and all covariates in the model by w_j. We then perform the linear regression using the weighted data, excluding the constant or intercept term from the regression and including a variable in the model whose values are w_j. This analysis is equivalent to the weighted analysis described in Section 4.4. In this situation we may calculate the logistic regression diagnostics using the equations given above for the situation when a case weight variable may be used. The only exception is the Pearson residual r_j which is:

$$r_j = \text{residual from the weighted least squares fit}$$

A number of different types of plots have been suggested for use, each directed at a particular aspect of fit. Some are formed from the seven diagnostics while others require additional computation. For example, see the methods based on grouping and smoothing in Landwehr, Pregibon, and Shoemaker (1984) and Fowlkes (1987). It is impractical to consider all possible suggested plots, so we will restrict attention to a few of the more easily computed ones which are meaningful in logistic regression analysis and may be considered to form the core of an analysis of diagnostics. These consist of the following:

(1) Plot ΔX_j^2 versus $\hat{\pi}_j$.

    (2) Plot ΔD_j versus $\hat{\pi}_j$.

    (3) Plot $\Delta \hat{\beta}_j$ versus $\hat{\pi}_j$.

Other plots that are sometimes useful include:

    (4) Plot ΔX_j^2 versus h_j.

    (5) Plot ΔD_j versus h_j.

    (6) Plot $\Delta \hat{\beta}_j$ versus h_j.

as these allow direct assessment of the contribution of leverage to the value of the diagnostic statistic. These plots, particularly (1)–(3), provide the basis for our analysis of diagnostics. Effective use of diagnostics is a good example of where there is no substitute for experience.

    As the plotting capabilities of microcomputer packages improve, additional plots will emerge. One we have found especially useful, which may be performed in SYSTAT's graphics module, is a plot of ΔX_j^2 versus $\hat{\pi}_j$ where the size of the plotting symbol is proportional in size to $\Delta \hat{\beta}_j$. This plot is illustrated in the examples which follow.

    To illustrate the use of the diagnostics and their related plots, we consider the final model for low birthweight given in Table 4.9. Recall that the summary statistics indicated that the model fit. Thus, we do not expect an analysis of diagnostics to show large numbers of covariate patterns being fit poorly. We might uncover a few covariate patterns which do not fit, or which have considerable influence on the estimated parameters. The key plots are given in Figures 5.3–5.7. We discuss each plot in turn.

    The diagnostics ΔX^2 and ΔD plotted versus the estimated logistic probabilities are shown in Figures 5.3 and 5.4. We prefer to use these plots instead of plots of r_j and d_j versus $\hat{\pi}_j$. The reasons for this choice are as follows: (1) When $J \approx n$, most positive residuals will correspond to covariate patterns where $y_j = m_j$, for example, 1, and negative residuals to those with $y_j = 0$. Hence, the sign of the residual is not useful. (2) Large residuals, irrespective of sign, correspond to poorly fit points. Squaring these residuals further emphasizes the lack of fit and removes the issue of sign. (3) The shape of the plot allows us to determine which patterns have $y_j = 0$ and which have $y_j = m_j$.

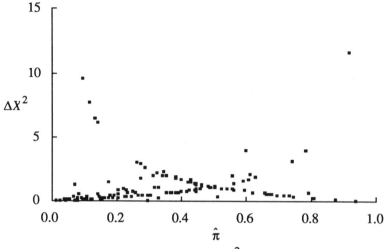

Figure 5.3 Plot of ΔX^2 Versus $\hat{\pi}$.

The shapes of the plots in Figures 5.3 and 5.4 are similar and show two quadratic like curves. The points on the curve going from the top left to bottom right corner correspond to covariate patterns with $y_j = m_j$. The abscissa for these points is proportional to $(1 - \hat{\pi}_j)^2$ since $m_j = 1$ for most covariate patterns. The points on the other curve, going from the bottom left to top right corner, correspond to covariate patterns with $y_j = 0$. The abscissa for these points is proportional to $(0 - \hat{\pi}_j)^2$. Covariate patterns that are poorly fit will generally be represented by points falling in the top left or top right corners of the plots. We will look for points that fall some distance from the balance of the data plotted. Assessment of "large" is part numeric value and part visual.

In Figure 5.3 we see 5 points, covariate patterns, which are poorly fit. These are four in the top left and one in the top right corner of the plot. These same five points may be easily seen in Figure 5.4. There are two other points with large values of ΔD which are found in the cup formed by the two quadratic like curves.

The range of ΔX^2 is much greater than ΔD. This is a property of Pearson versus deviance residuals. Whenever possible we prefer to use plots of both ΔX^2 and ΔD versus $\hat{\pi}$.

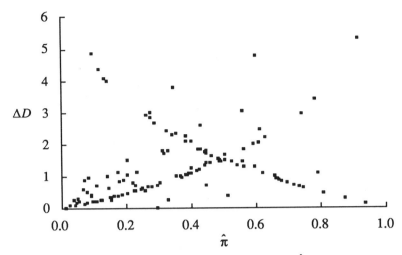

Figure 5.4 Plot of ΔD Versus $\hat{\pi}$.

Aside from the 5–7 points noted, the plots show that the model fits fairly well. Virtually all values of ΔX^2 and ΔD are less than 4. We use 4 as a crude approximation to the upper ninty-fifth percentile of the distribution of ΔX^2 and ΔD as, under m-asymptotics, these quantities would be distributed approximately as $\chi^2(1)$ and $\chi^2_{0.95}(1) = 3.84$.

The influence diagnostic $\Delta \hat{\beta}$ is plotted versus $\hat{\pi}$ in Figure 5.5.

We see two points with exceptionally large values and three others with moderately large values. Three of the five points fall well away from the balance of the data. As noted earlier, the largest values of $\Delta \hat{\beta}$ are likely to occur when both ΔX^2 and leverage are at least moderately large. The two points with the largest values of $\Delta \hat{\beta}$ correspond to patterns whose estimated logistic probabilities are in the interval where the highest leverage is likely to occur (see Table 5.1). The point with the third largest value corresponds to a pattern with one of the highest estimated logistic probabilities. For this point we suspect the major contribution to $\Delta \hat{\beta}$ is the ΔX^2 component since the leverage generally is quite small when $\hat{\pi} > 0.9$.

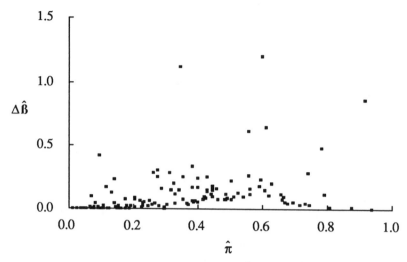

Figure 5.5 Plot of $\Delta\hat{\beta}$ Versus $\hat{\pi}$.

In Figure 5.6 we plot ΔX^2 versus $\hat{\pi}$ with the size of the symbol proportional $\Delta\hat{\beta}$). We have found that for values of $\Delta\hat{\beta}$ as small as 0.05 individual coefficients may exhibit substantial changes (i.e., greater than 20%). In order for these points to have a visual impact but not overwhelm the plot, multiplying $\Delta\hat{\beta}$ by 1.5 produces the best sized symbols in SYSTAT's graphics module. This plot allows us to clearly ascertain the contributions of residual and leverage to $\Delta\hat{\beta}$.

The two most influential covariate patterns are the two points with the largest circles. These points have moderate to large values of ΔX^2 and ΔD and have estimated probabilities in the region where highest leverage is expected. The point with the third largest value of $\Delta\hat{\beta}$ falls in the top right corner of the plots and, by its position, we know the major contribution to $\Delta\hat{\beta}$ comes from the large value of ΔX^2. In general the points in this plot which are of greatest concern will be those with large circles falling within the cup defined by the two quadratic curves. These will correspond to covariate patterns that are not fit very well and have high leverage.

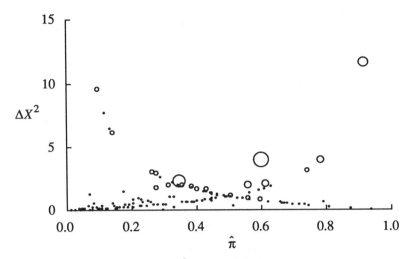

Figure 5.6 Plot of ΔX^2 Versus $\hat{\pi}$ with the Plotting Symbol
Proportional in Size to $1.5 \times \Delta\hat{\beta}$.

One problem with the influence diagnostic $\Delta\hat{\beta}$ is that it is a summary measure of change over all coefficients in the model simultaneously. For this reason it will be important to examine the changes in the individual coefficients due to specific covariate patterns identified as influential.

Examination of Figures 5.3–5.6 identifies four covariate patterns with exceptionally large values on one or more of the diagnostics statistics. These include two patterns with very large values of ΔX^2 and ΔD, and two more with moderate values of ΔX^2 and ΔD and very large $\Delta\hat{\beta}$. Information on these patterns is presented in Table 5.4. The quantity $P\#$ in Table 5.4 refers to the covariate pattern number. This number is somewhat arbitrary, as its value depends on how the data were aggregated. It should be noted that $P\#$ is not the original study ID.

The results in Table 5.4 provide an excellent example of what can be learned about a fitted model through diagnostic statistics. Consider covariate patterns 20 and 29. If either of these covariate patterns were deleted from the data set, a substantial change in the estimated coefficients would result, as would a decrease in the goodness-of-fit statistics X^2 and D. This is due to the fact that these patterns have high leverage and moderate values of fit reflected by ΔX^2 and

ΔD. In this case the diagnostics agree with the expectations set out in Table 5.3. For example, the $m = 3$ subjects with pattern number 20 each have an estimated logistic probability of 0.35, yielding a fitted value of $3 \times 0.35 = 1.05$. None of these 3 subjects had a low birth weight baby, resulting in an observed outcome of $y = 0$. Thus, pattern 20, characterized as young, white, smoking, low weight women with uterine irritability, is a configuration of covariates which has a high leverage and is not well fit by the model. The same is true for covariate pattern 29.

Table 5.4   Data, Observed Outcome (y), Number (m), Estimated Logistic Probability ($\hat{\pi}$), and the Value of the Four Diagnostic Statistics $\Delta\hat{\beta}$, ΔX^2, ΔD, and h for the Four Most Extreme Covariate Patterns ($P\#$).

P#	AGE	RACE	SMOKE	HT	UI	PTD	LWD	y	m	$\hat{\pi}$	$\Delta\hat{\beta}$	ΔX^2	ΔD^2	h
20	18	1	1	0	1	0	1	0	3	0.35	1.12	2.37	3.79	0.32
29	19	1	1	1	0	0	0	0	2	0.60	1.21	3.92	4.79	0.23
105	29	1	0	0	1	0	0	1	1	0.10	0.42	9.59	4.84	0.04
111	30	3	0	0	1	1	1	0	1	0.92	0.86	11.61	5.31	0.07

The results for patterns 105 and 111 portray a different type of outcome seen through diagnostic statistics. These patterns would not be expected to induce major changes in the estimated coefficients if removed from the data. This observation is also in agreement with the expectations in Table 5.3; the estimated probabilities for these two patterns are <0.1 or >0.9 (i.e., small h, and large ΔX^2). For example, the one individual with pattern number 111 presents a covariate pattern that has a very high estimated probability, 0.92, but the observed outcome is $y = 0$. This particular woman defied the model and produced a normal weight baby when her configuration of covariates would have suggested otherwise. The same thing can be said about pattern number 105, thought the relationship of the estimated probability and observed outcome is reversed.

We see two types of patterns: high leverage and poor fit (#'s 20 and 29) and low leverage and poor fit (#'s 105 and 111). How we handle these depends on the goals of the analysis. There appears to be little justification for excluding patterns 105 and 111 since the configuration of the covariates is biologically plausible. The observed outcomes are just unusual events. Exclusion of these patterns would not change values of the coefficients in the fitted model and thus not alter conclusions based upon it. Patterns 20 and 29 present a different challenge. If the goal of a study is risk factor identification, then we must examine carefully any single pattern that exerts a significant influence on the model. The particular pattern may represent an important new finding, in which case the pattern should be kept in the model. On the other hand, the configuration of covariates may be of little biologic interest.

Suppose that our goal is risk factor identification. We show in Table 5.5 an analysis of the effect of excluding covariate patterns 20 and 29 from the data. The effect is assessed in terms of the percent change in the estimated coefficients from the full data set.

Table 5.5 Estimated Coefficients from All Data, Estimated Coefficients Obtained by Deleting Selected Covariate Patterns, the Percent Change, and Values of Goodness-of-Fit Statistics for Eeach Model.

Variable	All Data	Delete 20	% Change	Delete 29	% Change	Delete 20 and 29	% Change
AGE	−0.08	−0.08	−3.4	−0.10	15.6	−0.09	12.2
RACE (1)	1.08	1.02	−6.2	0.96	−11.5	0.89	−18.0
RACE (2)	0.76	0.65	−14.2	0.70	−7.8	0.59	−22.9
SMOKE	1.15	1.14	−1.0	1.28	10.8	1.27	10.1
HT	1.33	1.39	4.7	2.00	50.3	2.04	53.3
UI	0.73	0.97	32.9	0.73	0.6	0.98	35.1
PTD	1.23	1.14	−7.2	1.20	−2.4	1.10	−10.1
LWD	−1.73	−0.85	−51.0	−1.93	11.6	−1.03	−40.7
A×L	0.15	0.11	−27.2	0.16	7.1	0.12	−20.8
S×L	−1.41		−22.5	−1.54	9.2	−1.22	−13.5
Constant	−0.51		6.1	−0.22	−56.3	−0.25	−50.4
Goodness-of-fit							
D	150.6	147.3		146.0		142.5	
X^2	137.4	132.1		136.7		131.3	
$\hat{C}$	5.2	6.6		3.9		5.4	

The results in Table 5.5 show that deletion of the two covariate patterns (5 of the 189 subjects) substantially alters the values of the estimated coefficients. These changes would alter the estimated odds ratios previously reported in Tables

4.10 and 4.12. Their deletion would also alter our opinion of which variables are needed in the final model. We would question the value of a variable selected for a model based on only one or two covariate patterns. It should be noted that, while not shown here, deletion of the covariate patterns did not alter the estimated standard errors for the estimated coefficients in any meaningful way. If we assume that the goal is risk factor identification, then we should seriously consider the importance of these two patterns. However, we should not exclude data solely on statistical grounds, appropriate consultation with clinicians should take place, and exclusions made with appropriate attention to clinical issues. An alternative to exclusion of the patterns is to try to collect more data from subjects with these same covariate patterns, refit the model, and hope the new model is less sensitive to this region of the covariate space.

The model for the low birth weight data is an example where the model fits well, and use of diagnostics identified only a few covariate patterns where the model did not fit, and/or the patterns were influential. Suppose instead that we have a model where the summary statistics indicate that there is substantial deviation from fit. In this situation, we have evidence that for more than a few covariate patterns, y_j differs from $m_j\hat{\pi}_j$. One or more of three things has likely happened: (1) the logistic model does not provide a good approximation to the correct relationship between the conditional mean, $E(Y \mid \mathbf{x}_j)$, and $\mathbf{x}_j$, (2) we have not measured and/or not included an important covariate into the model, or (3) one or more covariates in the model have not been entered in the correct scale. We will discuss each of these in turn.

The logistic regression model is a remarkably flexible model. Unless we are dealing with a set of data where most of the probabilities are very small or very large, or where the fit is extremely poor in an identifiable systematic manner, it is unlikely that any alternative model will provide a better fit. Cox (1970) demonstrates that the logistic and other, similar symmetric models are virtually identical in the region from 0.2 to 0.8. If one suspects, based on biologic or other reasons, such as graphical presentations, or the "goodness-of-link tests" proposed by Pregibon (1980), that the logistic model is the wrong one, then careful thought should be given to the choice of the alternative model. Particular attention should be given to issues of interpretation. Are the coefficients biologically interpretable? The approach that tries all other possible models and selects the "best fitting" one is not recommended, as no thought is given to the biological implications of the selected model. In some situations, inadequacy of a fitted logistic model can be corrected by addressing the issues of variable selection and scale identification. However, we must keep in mind the distinction between getting a model to fit and having the theoretically correct model.

Some interesting theoretical work has been done by White (1982, 1989) and Hjort (1988) on the use of maximum likelihood estimation with a misspecified model. These authors show that the fitted logistic regression model is the one that minimizes the Kullbeck–Leibler information distance between the theoretically correct model and the logistic model. In this sense the fitted logistic regression model is a best approximation to the true model. White provides a test for the hypothesis that the fitted model is the theoretically correct one. The test is elegant but is difficult to compute in practice and its power has not been adequately studied. Hence, we recommend that assessment of the adequacy of the fitted logistic model be performed using the methods suggested in this chapter. When there is evidence that the logistic model does not fit the data an alternative model should be selected on the basis of biological consideration.

From the point of view of variable selection the analysis reverts back to what are really model-building techniques. Model fitting is an iterative procedure. We rarely obtain a the final model on the first pass through the data.

We hope that the study was carefully enough designed so no major covariates were excluded. However, it is possible that the biological factors associated with the outcome variable are not well known and in this case a key variable may not be present in the observed data. Little can be done if this is the case, except to go back and collect it, if it is now known. The potential biases and pitfalls of this statistical oversight are enormous. This approach of retroactive data collection is also impractical in most research situations.

Another possible cause of lack of fit is for the variability in the outcome variable to exceed what would be predicted by the model and binomial variation. A commonly cited example comes from toxicological experiments where pups in litters of the same parentage are observed [see Haseman and Hogan (1975), Haseman and Kupper (1979), and Williams (1975)]. This source of lack of fit is often called extrabinomial variation. Several different models have been proposed as a method for incorporating it into logistic regression [see Crowder (1978), Pierce and Sands (1975), Mauritsen (1984), and Williams (1982)]. The basic approach is to incorporate additional parameters into the model. Their exact form and nature depend on which model is used. The idea is similar to incorporating random effects into an ANOVA model. The EGRET package provides an option for testing and modeling for this type of variability. A constraint on the method in EGRET is that m must be ≥ 2 for most covariate patterns. A method for modeling extra sources of variation in GLIM may be found in Williams (1982). When incorporating these models into an analysis we must be cognizant of their biologic plausibility and avoid inclusion of meaningless parameters merely to improve the fit.

The third possibility is that the methods initially used for variable identification and scale selection were not precise enough to accomplish the task. A number of graphical methods have recently been proposed which show some promise. The generalized additive models approach of Hastie and Tibshirani (1986, 1987) has already been discussed in Chapter 4. We may choose to employ their methods to help us define the correct scale for potential variables. In linear regression, one technique available to the analyst is to use partial residuals to help identify the correct scale of a variable. Fowlkes (1987) has extended Landwehr, Pregibon, and Shoemaker's (1984) method of partial residuals for logistic regression into a form that eliminates some of the structure due to the binary nature of the response variable. These methods are computationally intensive and are difficult to carry out without special purpose software. As software is further developed these methods may become options available in packages for logistic regression.

Once new or alternative scales have been chosen, the new model is then subjected to the methods for assessment of fit. This process continues until we are satisfied with the model.

5.4 Assessment of Fit via External Validation

In some situations it may be possible to exclude a subsample of our observations, develop a model based on the remaining subjects, and then test the model in the originally excluded subjects. In other situations it may be possible to obtain a new sample of data to assess the goodness-of-fit of a previously developed model. This type of assessment is often called model validation, and may be especially important when the fitted model is used to predict outcome for future subjects. The reason for considering this type of assessment of model performance is that the fitted model always performs in an optimistic manner on the developmental data set. This has been a major concern to users of discriminant analysis, and considerable work has been done in this area [see, e.g., Efron (1986)]. The use of validation data amounts to an assessment of goodness-of-fit where the fitted model is considered to be theoretically known, and no estimation is performed. Some of the diagnostics discussed in Section 5.3 (ΔX^2, ΔD, $\Delta \hat{\beta}$) mimic this idea by computing, for each covariate pattern, a quantity based on the exclusion of the particular covariate pattern. With a new data set a more thorough assessment is possible.

The methods for assessment of fit in the validation sample parallel those described in Sections 5.2 and 5.3 for the developmental sample. The major difference is that the values of the coefficients in the model are regarded as fixed constants rather than estimated values.

Suppose that the validation sample consists of n_v observations (y_i, x_i), $i = 1, 2, ..., n_v$, which may be grouped into J_v covariate patterns. In keeping with previous notation, let y_j denote the number of positive responses among the m_j subjects with covariate pattern $x = x_j$ for $j = 1, 2, ..., J_v$. The logistic probability for the jth covariate pattern is π_j, the value of the previously estimated logistic model using the covariate pattern, x_j, from the validation sample. These quantities become the basis for the computation of the summary measures of fit, X^2, D, and C, from the validation sample. Each of these will be considered in turn.

The computation of the Pearson chi-square statistic follows directly from equation (5.2), with obvious substitution of quantities from the validation sample. In this case X^2 is computed as the sum of J_v independent terms. If each $m_j\pi_j$ is large enough to use the normal approximation to the binomial distribution, then X^2 will be distributed, under the hypothesis that the model is correct, as $\chi^2(J_v)$. We expect that in practice the observed numbers of subjects within each covariate pattern will be small, with most $m_j = 1$. Hence, m-asymptotics will not be able to be employed. In this case about the best we can do is compare X^2 to its expectation, J_v. Similar considerations apply to the calculation and interpretation of D for the validation sample. The model is considered to provide a poor fit to the validation data if X^2 and D exceed J_v by a substantial amount. How much is too much is a difficult thing to decide, and will depend on experience and the overall picture we get from all the methods used to check the fit in the validation sample.

The same line of reasoning discussed in Section 5.2.2 to develop the Hosmer–Lemeshow test may be used to obtain an equivalent statistic for the validation sample. Suppose for sake of simplicity of presentation, we assume that we wish to use 10 groups composed of the deciles of risk. Any other grouping strategy could be used with obvious modifications in the calculations. Let n_k denote the approximately $n_v/10$ subjects in the kth decile of risk. Let $o_k = \Sigma m_j y_j$ be the number of positive responses among the covariate patterns falling in the kth decile of risk. The estimate of the expected value of o_k under the assumption that the model is correct is $e_k = \Sigma m_j\pi_j$, where the sum is over the covariate patterns in the decile of risk. The Hosmer–Lemeshow statistic is obtained as the Pearson chi-square statistic computed from the observed and expected frequencies

$$C_v = \sum_{k=1}^{g} \frac{(o_k - e_k)^2}{n_k \bar{\pi}_k (1 - \bar{\pi}_k)} \tag{5.13}$$

where $\bar{\pi}_k = \Sigma m_j\pi_j/n_k$. The subscript, v, has been added to C_v to emphasize that the statistic has been calculated from a validation sample. Under the hypothesis that the model is correct, and the assumption that each e_k is sufficiently large for each term in C_v to be distributed as $\chi^2(1)$ it follows that C_v is distributed as $\chi^2(10)$. In general, if we use g groups then the distribution will be $\chi^2(g)$.

In addition to calculating a p-value to assess overall fit, we recommend that each term in C_v be examined to assess the fit within each decile of risk. The comments given in Section 5.2.2 regarding modification of the denominator of the test statistic, $\hat{C}$, in equation (5.5) also apply to C_v equation (5.13).

The classification table is the remaining summary statistic that we are likely to use with the validation sample, and then only in instances where classification is an important use of the model. The classification table is obtained in exactly the same manner as shown in Section 5.2.3, with the modification that probabilities are no longer thought of as being estimated. The resulting table may then be used to compute statistics such as sensitivity, specificity, positive, and negative predictive power. Interpretation of these will depend on the particular situation.

Exercises

1. As is the case in linear regression effective use of diagnostic statistics depends on our ability to interpret and understand the values of the statistics. The purpose of this problem is to provide a few structured examples to examine the effect on the fitted logistic regression model and diagnostic statistics when data are moved away from the model (i.e., poorer fit), and also toward the model (i.e., better fit). Listed in the table below is a column of values of the independent variable, x, and seven different columns of outcome variable, y, labeled "model." All models fit in this problem use the given values of x for the covariate. Different models will be fit using the different columns for the outcome variable. The data for the column labeled "model 0" are a subset of the data used to generate Figure 5.1. Thus, these data represent a "typical" realization when the logistic regression model is correct. In the columns labeled " model 1" to "model 3" we have changed some of the y values away from the original model. Namely some cases with small values of x have had y changed from 0 to 1 and others with large values of x have had the y values changed to 1. For models labeled "model −1" and "model −2" we have moved the y values in the direction of the model. That is, we have changed y from 1 to 0 for some small values of x and have changed y from 0 to 1 for some

large values of x. Fit the six logistic regression models columns "model –
2" to "model 3." Compute for each fitted model the values of the leverage,
h, the change in chi-square, ΔX^2, and the influence diagnostic, $\Delta \hat{\beta}$. Plot
each of these versus the fitted values, predicted logistic probabilities.
Compare the plots over the various models. Do the statistics pick out
poorly fit and influential cases? How do the estimated coefficients change
relative to model 0? Fit "model $-i$." What happens and why? Recall the
discussion in Section 4.5 on complete separation.

	Model						
x	$-i$	-2	-1	0	1	2	3
−5.65	0	0	0	0	0	0	0
−4.75	0	0	0	0	0	1	1
−3.89	0	0	0	0	0	0	0
−3.12	0	0	0	0	0	0	0
−2.93	0	0	0	0	0	0	0
−2.87	0	0	0	0	0	0	0
−1.85	0	0	0	0	1	1	1
−1.25	0	1	1	1	1	1	1
−0.97	0	0	0	0	0	0	0
−0.19	1	1	1	1	1	1	1
−0.15	1	1	1	1	1	1	1
0.69	1	1	1	1	1	1	1
1.07	1	1	1	1	1	1	1
1.18	1	1	1	1	1	1	1
1.45	1	1	0	0	0	0	0
2.33	1	1	1	0	0	0	0
3.57	1	1	1	1	1	1	1
4.41	1	1	1	1	1	1	1
4.57	1	1	1	1	1	1	0
5.85	1	1	1	1	1	1	1

2. In the exercises in Chapter 4, problem 2, multivariate models for the ICU
    study were formed. Assess the fit of the model(s) that you feel was (were)
    best among those considered. This assessment should include an overall
    assessment of fit and use of the diagnostic statistics. Does the model fit?

Are there any particular subjects, covariate patterns, which seem to be poorly fit or overly influential? If so, how would you propose to deal with them?

CHAPTER 6

Application of Logistic Regression with Different Sampling Models

6.1 Introduction

The material in this chapter is required to establish the likelihood functions used in later chapters. The mathematical content is at a higher level than other chapters. Mathematically less sophisticated readers may skip this chapter without compromising their understanding of the methods presented in later chapters.

Up to this point we have assumed that the data have come from a simple random sampling scheme which will allow for estimation of the conditional probability, $P(Y \mid \mathbf{x})$, through the use of the likelihood function shown in equation (1.2). At issue in this chapter is not the appropriateness of the logistic regression model to describe the conditional probability. The focus is instead the study design and construction of the appropriate likelihood function. We will consider each of the more common types of study design in which logistic regression is used.

Since the goal is to provide an overview of the issues, technical discussion will be kept to a minimum. References to the literature for more detailed treatment will be provided. Throughout this chapter we will assume that the outcome variable is dichotomous, coded as zero or 1, and that its conditional probability given a vector of covariates is the logistic regression model. In addition we assume that the number of covariate patterns is equal to the sample size. Modifications to allow for replication at covariate patterns are a notational detail, not a conceptual problem.

6.2 Cohort Studies

Several variations of the cohort (or prospective) study are in common use. The simplest design is where a simple random sample of subjects is chosen and the values of the covariates are determined. These subjects are then followed for a fixed period of time and the outcome variable is measured. This type of sample is identical to what is often referred to as the regression sampling model, in which we assume that the values of the covariates are fixed and measured without error and the outcome is measured conditionally on the observed values of the covariates. Under these assumptions and independence of the observations, the likelihood function for a sample of size n is

$$L_1(\beta) = \prod_{i=1}^{n} P(Y_i = y_i \mid \mathbf{x}_i) \qquad (6.1)$$

When the observed values of y and the logistic regression model are substituted into the expression for the conditional probability, $L_1(\beta)$ simplifies to the likelihood function in equation (1.2).

A modification of this situation is that of a randomized trial where subjects are first chosen via a simple random sample and then allocated independently and with known probabilities into "treatment" groups. Subjects are followed and the outcome variable measured. If the responses are such that a normal errors model is appropriate we would be naturally led to consider a normal theory analysis of covariance model which would contain appropriate design variables for treatment, relevant covariates, and any interactions between treatment and covariates which are deemed necessary. The extension of the likelihood function in equation (6.1) to incorporate treatment and covariate information when the outcome is dichotomous is obtained by including these variables in the logistic regression model.

Another modification is for the design to incorporate a stratification variable such as location or clinic. In this situation the likelihood function is the product of the stratum specific likelihood function each of which is similar in form to $L_1(\beta)$. We would perhaps add terms to the model to account for stratum specific responses. These might include a design variable for stratum and interactions between this design variable and other covariates.

In each of these designs we use the likelihood function $L_1(\beta)$ as a basis for determining the maximum likelihood estimates of the unknown parameters in the vector β. Tests and confidence intervals for the parameters follow from well-developed theory for maximum likelihood estimation [see Cox and Hinkley (1974)]. The estimated parameters may be used in the logistic regression model to estimate the conditional probability of response for each subject. The fact that the estimated logistic probability provides a model-based estimate of the probability of response permits the development of methods for assessment of goodness-of-fit such as those discussed in Chapter 5. Chambless and Boyle (1985) have extended $L_1(\beta)$ to the case where the data come from a stratified simple random sample.

A noticeable gap exists which extends the use of logistic regression into the domain of more complex sampling plans such as cluster and multistage sampling. However, the necessary theory is well developed for normal theory regression analyses of studies using complex sampling plans. Given the similarity between normally distributed errors in linear regression analyses and binomial errors in logistic regression, the extension of existing methodology for logistic regression to handle complex sampling plans should not be difficult.

In some prospective studies the outcome variable of interest is the time to the occurrence of some event. In these studies the time to event is often modeled using the proportional hazards model [see Cox and Oakes (1984), Kalbfleisch and Prentice (1980), or Lee (1980)]. In these situations a method of analysis which is sometimes used is to ignore the actual failure time and model the occurrence or nonoccurrence of the event via logistic regression Because this mode of analysis is frequently used we discuss it in detail in Chapter 8.

6.3 Case-Control Studies

One of the major reasons the logistic regression model has seen such wide use, especially in epidemiologic research, is the ease of obtaining adjusted odds ratios from the estimated slope coefficients when sampling is performed conditional on the outcome variables, as in a case-control study. Cornfield

(1951) is generally given credit for first observing that the odds ratio is invariant under study design, cohort, or case-control. However, it was not until the work of Farewell (1979) and Prentice and Pyke (1979) that the mathematical details justifying the common practice of analyzing case-control data as if they were cohort data were worked out.

In contrast to cohort studies, the binary outcome variable in a case-control study is fixed by stratification. The dependent variables in this setting are one or more primary covariates, exposure variables in **x**. In this type of study design samples of fixed size are chosen from the two strata defined by the outcome variable. The values of the primary exposure variables and the relevant covariates are then measured for each subject selected. The covariates are assumed to include all relevant exposure, confounding, and interaction terms. The likelihood function is the product of the stratum-specific likelihood functions which depends on the probability that the subject was selected for the sample, and the probability distribution of the covariates.

It is not difficult to algebraically manipulate the case-control likelihood function to obtain a logistic regression model in which the dependent variable is the outcome variable of interest to the investigator. The key steps in this development are two applications of Bayes theorem. Since the likelihood function will be based on subjects selected, we need to define a variable that records selection status for each subject in the population. Let the variable s denote the selection, $s = 1$, or nonselection, $s = 0$, of a subject. The full likelihood for a sample of size n_1 cases ($y = 1$) and n_0 controls ($y = 0$) is

$$\prod_{i=1}^{n_1} P(\mathbf{x}_i \mid y_i = 1, s_i = 1) \prod_{i=1}^{n_0} P(\mathbf{x}_i \mid y_i = 0, s_i = 1) \tag{6.2}$$

For an individual term in the likelihood function shown in equation (6.2) the first application of Bayes theorem yields

$$P(\mathbf{x} \mid y, s = 1) = \frac{P(y \mid \mathbf{x}, s = 1)\, P(\mathbf{x} \mid s = 1)}{P(y \mid s = 1)} \tag{6.3}$$

The second application of Bayes theorem is to the first term in the numerator of equation (6.3). This yields, when $y = 1$,

$$P(y = 1 | \mathbf{x}, s = 1)$$

$$= \frac{P(y = 1 | \mathbf{x}) \, P(s = 1 | \mathbf{x}, y = 1)}{P(y = 0 | \mathbf{x}) \, P(s = 1 | \mathbf{x}, y = 0) \; P(y = 1 | \mathbf{x}) \, P(s = 1 | \mathbf{x}, y = 1)}$$

$$(6.4)$$

Assume that the selection of cases and controls is independent of the covariates with respective probabilities τ_1 and τ_0 ; then

$$\tau_1 = P(s = 1 | y = 1, \mathbf{x})$$

$$= P(s = 1 | y = 1)$$

and

$$\tau_0 = P(s = 1 | y = 0, \mathbf{x})$$

$$= P(s = 1 | y = 0)$$

Substitution of τ_1, τ_0 and the logistic regression model, $\pi(\mathbf{x})$, for $P(y = 1 | \mathbf{x})$, into equation (6.4) yields

$$P(y = 1 | \mathbf{x}, s = 1) = \frac{\tau_1 \pi(\mathbf{x})}{\tau_0 \, [1 - \pi(\mathbf{x})] + \tau_1 \pi(\mathbf{x})} \qquad (6.5)$$

If we divide the numerator and denominator of the expression on the right hand side of equation (6.5) by $\tau_0[1 - \pi(\mathbf{x})]$, the result is a logistic regression model with intercept term $\beta_0^* = \ln(\tau_1/\tau_0) + \beta_0$. To simplify the notation, let $\pi^*(\mathbf{x})$ denote the right-hand side of equation (6.5). By assumption sampling is independent of covariate values, thus, $P(\mathbf{x} | s = 1) = P(\mathbf{x})$ where $P(\mathbf{x})$ denotes the probability distribution of the covariates. The general term in the likelihood shown in equation (6.3) then becomes, for $y = 1$,

$$P(\mathbf{x} | y = 1, s = 1) = \frac{\pi^*(\mathbf{x}) P(\mathbf{x})}{P(y = 1 | s = 1)} \qquad (6.6)$$

A similar term for $y = 0$ is obtained by replacing $\pi^*(\mathbf{x})$ by $[1 - \pi^*(\mathbf{x})]$ in the numerator and $P(y = 1 | s = 1)$ by $P(y = 0 | s = 1)$ in the denominator of equation (6.6). If we let

$$L^*(\beta) = \prod_{i=1}^{n} \pi^*(\mathbf{x}_i)^{y_i} [1 - \pi^*(\mathbf{x}_i)]^{1-y_i}$$

then the likelihood function shown in equation (6.2) becomes

$$L^*(\beta) \prod_{i=1}^{n} \left[\frac{P(\mathbf{x}_i)}{P(y_i \mid s_i = 1)} \right] \tag{6.7}$$

The first term in equation (6.7), $L^*(\beta)$, is the likelihood obtained when we pretend the case-control data were collected in a cohort study, with the outcome of interest modeled as the dependent variable. If we assume that the probability distribution of $\mathbf{x}$, $P(\mathbf{x})$, contains no information about the coefficients in the logistic regression model then maximization of the full likelihood with respect to the parameters in the logistic model $\pi^*(\mathbf{x})$ is only subject to the restriction that $P(y_i = 1 \mid s_i = 1) = n_1/n$ and $P(y_i = 0 \mid s_i = 1) = n_0/n$. The likelihood equation obtained by differentiating with respect to the parameter β_0^* assures that this condition will be satisfied. Thus, maximization of the full likelihood with respect to the parameters in $\pi^*(\mathbf{x})$ need only consider that portion of the likelihood which looks like a cohort study. The implication of this is that *analysis of data from case-control studies via logistic regression may proceed in the same way and using the same computer programs as cohort studies.* Nevertheless, inferences about the intercept parameter β_0 are not possible without knowledge of the sampling fractions, within cases and controls, τ_1, τ_0.

The assumption that the marginal distribution of $\mathbf{x}$ contains no information about the parameters in the logistic regression model requires additional discussion as it is not true in one historically important situation, the normal theory discriminant function model. This model was discussed briefly in Chapters 1 and 2. When the assumptions for the normal discriminant function model hold the maximum likelihood estimators of the coefficients for the logistic regression model obtained from conditional likelihoods such as those in equations (6.2) and (6.7) will be less efficient than the discriminant function estimator shown in equation (2.7) [see Efron (1975)]. However, the assumptions for the normal theory discriminant function model are rarely, if ever, attained in practice. Application of the normal discriminant function when its assumptions do not hold may result in substantial bias, especially when some of the covariates are dichotomous variables. As a general rule, estimation should be based on equations (6.2) and (6.7), unless there is considerable evidence in favor of the normal theory discriminant function model.

Prentice and Pyke (1979) have shown that the maximum likelihood estimators obtained by pretending that the case-control data was a cohort sample have the usual properties associated with maximum likelihood estimators. Specifically, they will be asymptotically normally distributed, with covariance matrix obtained from the inverse of the information matrix. Thus, percentiles from the $N(0,1)$ distribution may be used in conjunction with estimated standard errors produced from standard logistic regression software to form Wald statistics and confidence interval estimates. The theory of likelihood ratio tests may be employed to compare models obtained via the difference in the deviance of the two models, assuming of course that the models are nested. Scott and Wild (1986) have shown that inferences based on this approach are sensitive to having the correct logit function. They show that failure to include necessary higher order terms in the logit will produce a model with estimated standard errors that are too small. These results are special cases of more general results obtained by White (1982).

Modification of the likelihood function to incorporate additional levels of stratification beyond case-control status follows in the same manner as described for cohort data, for example, inclusion of relevant design variables and interaction terms. Thus, model building and inferences from fitted models for case-control data may proceed using the methods developed for cohort data, as described in Chapters 4 and 5. However, this approach is not valid for matched or highly stratified data. Appropriate methods for the analysis of the latter are presented in detail in Chapter 7.

Fears and Brown (1986) have described an alternative method for the analysis of stratified case-control data which does not require the inclusion of stratum-specific terms. Their model does require that we know the sampling rates and the total number of subjects in each stratum. This information is then used to define the relative sampling rates for cases and controls within each stratum. The ratio of these is included in the model in the form of an additional known constant added to the stratum specific logit. Specifically, suppose we let n_j be the total number of subjects with $y = j$ observed out of a possible N_j and let the ith stratum-specific quantities be n_{jk} and N_{jk}, $j = 0$, 1 and $k = 1, 2 ..., K$. The relative stratum-specific sampling rates are $w_{1k} = (n_{1k}/N_{1k})/(n_1/N_1)$ and $w_{0k} = (n_{0k}/N_{0k})/(n_0/N_0)$. The Fears and Brown model uses stratum-specific logits of $g_k(\mathbf{x}) = \ln(w_{1k}/w_{0k}) + \beta_0 + \text{ß}'\mathbf{x}$, $k = 1, 2, ..., K$. This model may be handled with standard logistic regression software by defining a new variable which takes on the value $1 + \ln(w_{1k}/w_{0k})$ and forcing it into the model while simultaneously deleting the usual constant term. Fears and Brown point out that with this model we may include, as covariates, variables that define strata. This implies modeling them as continuous, because, to include them in design variable form is equivalent to inclusion of individual stratum-specific parameters that Fears and Brown want to avoid.

Breslow and Cain (1988) show that the estimator proposed by Brown and Fears is asymptotically normally distributed and is the conditional maximum likelihood estimator but not the restricted maximum likelihood estimator as claimed by Brown and Fears. Breslow and Zaho (1988) point out that the estimated standard errors produced when standard logistic regression software is used to implement the Brown and Fears method overestimates the true standard errors. They provide an expression for a covariance matrix which will yield consistent estimates of the standard error. The matrix is complicated to compute as it requires a special purpose program or a high degree of skill in using a flexible package such as GLIM. For these reasons we do not present the Breslow and Zhao estimator in detail.

Before leaving our discussion of logistic regression in the case-control setting, we briefly consider the application of the chi-square goodness-of-fit tests for the logistic regression model presented in Section 5.2. The essential feature of these tests is that for a particular covariate pattern, the number of subjects with the outcome response of interest responding among m sampled is distributed binomially with parameters m and response probability given by the hypothesized logistic regression model. Recall that for cohort data, the likelihood function was parameterized directly in terms of the logistic probability. For case-control data, the function $\pi^*(\mathbf{x})$ is the probability $P(y = 1 | \mathbf{x}, s = 1)$. For a particular covariate pattern, conditioning on the number of subjects m observed to have a given covariate pattern is equivalent to conditioning on the event, $(\mathbf{x}, s = 1)$. Thus, for case-control studies in which the logistic regression model assumption is correct, the conditional distribution of the number of subjects responding among the m observed to have a particular covariate pattern is binomial with parameters m and $\pi^*(\mathbf{x})$. Hence, the results developed in Chapter 5 based on m-asymptotics also apply.

It is often the case that data from case-control studies do not arise from simple random samples within each stratum. For example, the design may call for the inclusion of all subjects with $y = 1$ and a sample of subjects with $y = 0$. For these designs there is an obvious dependency among the observations. If this dependency is not too great, or if we appeal to a superpopulation model [see Prentice (1986)], then employing a theory ignoring it should not bias the results significantly.

6.4 Other Sampling Models

A new development by Steinberg and Cardell (1987) and related work by Cardell and Steinberg (1987) permits application of logistic regression to a setting where data are available for a sample of subjects who have $y = 1$ and a second sample where the value of y is unknown but the values of the covariates are known. In particular we assume that the first sample is a simple random sample of size n_1 chosen from among the N_1 subjects with $y = 1$. We must know the value of N_1. The second sample is a simple random sample of size n from the total population of size N, N known, containing an unknown number of subjects with $y = 1$ and $y = 0$. Consistent estimates of the parameters of the logistic regression model may be obtained from logistic regression software provided the software allows for negative weights. The method requires that we construct a data set containing $n + 2n_1$ observations. The first n observations consist of the covariates for the n observations in the second sample with the value of the outcome variable $y = 0$ and the weight $w = 1$. The next n_1 observations consist of the covariates for the n_1 subjects in the first sample with the outcome variable $y = 1$ and the weight $w = (n/N)/(n_1/N_1)$. The last n_1

observations contain a repeat of the covariates for the n_1 subjects in the first sample but with the outcome variable $y = 0$ and weight $w = -(n/N)/(n_1/N_1)$.

We obtain the sample estimate of the covariance matrix of the estimated parameters via matrix calculations using only the n observations in the second sample. Let **X** be the n by $p + 1$ matrix containing the covariate values for these subjects and let **V** denote the n by n diagonal matrix with arbitrary element $v_i = \hat{\pi}(x_i)[1 - \hat{\pi}(x_i)]$, where $\hat{\pi}(x_i)$ is the estimated logistic probability for the ith subject in this sample. This estimated logistic probability uses the parameter estimates obtained from the data set containing the $n + 2n_1$ observations just described. Let **T** be the n by n diagonal matrix with general element $t_i = 2(1 - n/N)\hat{\pi}(x_i)^2 + (w - 1)\hat{\pi}(x_i)$ where $w = (n/N)/(n_1/N_1)$. The estimator of the covariance matrix of the maximum likelihood estimators is

$$(\mathbf{X'VX})^{-1} + (\mathbf{X'VX})^{-1}(\mathbf{X'TX})(\mathbf{X'VX})^{-1}$$

This formulation may be useful in fields in which $y = 1$ is called a choice based sample, referring to the fact that $y = 1$ may represent the choice of some product. The second sample of data may come from an existing data base or other source containing information on the covariates for the population in general. There is an obvious potential for a number of significant sources of bias in the data in such studies. If these are not present then the methods described should prove useful in describing the associations between covariates and outcome.

Exercises

1. Use the data presented in Breslow and Zaho (1988) to perform the analyses reported in that paper and the analysis reported in Fears and Brown (1986).

2. Consider the mammography experience data presented in Appendix 6. See Section 8.1 for a brief description of this study and a code sheet for the variables. Suppose we assume that the women with ME = 1 or 2 among the first 200 subjects (OBS $\leq$ 200 in Appendix 6) represent a choice based sample with outcome present, $y = 1$ from $N_1 = 10,000$ women who have had a mammogram. Suppose that subjects 201 through 412 (OBS >200 in

Appendix 6) represent a second sample drawn from a population of size $N =$ 30,000 women, and that we do not know which women have had a mammogram. Consider as covariates the variables perceived benefit, PB, and family history, HIST. Use the method of Steinberg and Cardell to fit a logistic regression model to these data.

CHAPTER 7

Logistic Regression for Matched Case-Control Studies

7.1 Introduction

An important special case of the stratified case-control study discussed in Chapter 6 is the matched study. A discussion of the rationale for matched studies may be found in epidemiology texts such as Breslow and Day (1980), Kleinbaum, Kupper, and Morgenstern (1982), Schlesselman (1982), Rothman (1986), and Kelsey, Thompson, and Evans (1986). In this study design subjects are stratified on the basis of variables believed to be associated with the outcome. Age and sex are examples of commonly used stratification variables. Within each stratum samples of cases ($y = 1$) and controls ($y = 0$) are chosen. The number of cases and controls need not be constant across strata, but the most common matched designs include one case and from one to five controls per stratum and are thus referred to as $1-M$ matched studies.

In this chapter we develop the methods for analysis for the general case. Greater detail is provided for the $1-1$ design, as it can be analyzed using standard logistic regression software. An example of a $1-3$ matched study is also provided.

We begin by providing some motivation and rationale for the need for special methods for the matched study. In Chapter 6 it was noted that we could handle the stratified sample by including the design variables created from the stratification variable in the model. This approach works well when the number subjects in each stratum is large. However, in matched studies we are likely to have few subjects per stratum. For example, in the $1-1$ matched design with n case-control pairs we have only two subjects per stratum. Thus, in a fully stratified analysis with p covariates, we would be required to estimate $(n - 1) + p$ parameters consisting of the p slope coefficients for the covariates and the $n - 1$

coefficients for the stratum-specific design variables using a sample of size $2n$. The optimality properties of the method of maximum likelihood, derived by letting the sample size become large, hold only when the number of parameters remains fixed. In any $1-M$ matched study this is clearly not the case. With the fully stratified analysis, the number of parameters increases at the same rate as the sample size. For example, with a model containing one dichotomous covariate it can be shown that the bias in the estimate of the coefficient is 100% when analyzing a matched $1-1$ design via a fully stratified likelihood. If we regard the stratum specific parameters as nuisance parameters and are willing to forgo their estimation, then we can create a conditional likelihood which will yield maximum likelihood estimators of the slope coefficients in the logistic regression model which are consistent and asymptotically normally distributed. The mathematical details of conditional likelihood analysis may be found in Cox and Hinkley (1974). We summarize its application to the matched design. Liang (1987), in related work, considers a general approach to the analysis of highly stratified data.

Suppose that there are K strata with n_{1k} cases and n_{0k} controls in stratum k, $k = 1, 2, ..., K$. The conditional likelihood for the kth stratum is obtained as the probability of the observed data conditional on the stratum total and the total number of cases observed, the sufficient statistic for the nuisance parameter. In this case it is the probability of the observed outcome relative to the probability of the data for all possible assignments of n_{1k} cases and n_{0k} controls to $n_k = n_{1k} + n_{0k}$ subjects. There are $\binom{n_k}{n_{1k}}$ possible assignments of case status to n_{1k} subjects among the n_k subjects in the stratum. Let the subscript j denote any one of these assignments. For any assignment we let subjects 1 to n_{1k} correspond to the cases and subjects $n_{1k} + 1$ to n_k to the controls. This will be indexed by i for the observed data and by i_j for the jth possible assignment. The conditional likelihood may be expressed as

$$l_k(\beta) = \frac{\prod_{i=1}^{n_{1k}} P(\mathbf{x}_i | y = 1) \prod_{i=n_{1k}+1}^{n_k} P(\mathbf{x}_i | y = 0)}{\sum_j \left\{ \prod_{i_j=1}^{n_{1k}} P(\mathbf{x}_{ji_j} | y = 1) \prod_{i_j=n_{1k}+1}^{n_k} P(\mathbf{x}_{ji_j} | y = 0) \right\}} \tag{7.1}$$

where the summation over j in the denominator is over the n_k choose n_{1k} combinations. The full conditional likelihood is the product of the $l_k(\beta)$ over the K strata,

$$l(\beta) = \prod_{k=1}^{K} l_k(\beta) \tag{7.2}$$

Let the logit in the kth stratum be $g_k(\mathbf{x}) = \alpha_k + \beta'\mathbf{x}$, where α_k denotes the contribution to the logit of all terms constant within the stratum, the stratification variable(s). In this chapter the vector of coefficients β contains only the p slope coefficients, $\beta' = (\beta_1, \beta_2, ..., \beta_p)$. Assuming that the logistic regression model is correct, application of Bayes theorem to each term in $l_k(\beta)$ in equation (7.1) yields

$$l_k(\beta) = \frac{\prod_{i=1}^{n_{1k}} e^{\beta'\mathbf{x}_i}}{\sum_J \prod_{i_j=1}^{n_{1k}} e^{\beta'\mathbf{x}_{ji_j}}} \tag{7.3}$$

Note that the terms of the form $e^{\alpha_k}/(1 + e^{\alpha_k + \beta'\mathbf{x}})$ appear equally in both the numerator and denominator of equation (7.1) and thus cancel out. This leaves the function shown in equation (7.3) which depends only on β. The conditional maximum likelihood estimator for β is that value which maximizes equation (7.2) when $l_k(\beta)$ is as shown in equation (7.3). We show in the next section that the 1–1 matched design may be analyzed using standard logistic regression software. Other software must be used when $M > 1$. The EGRET (1988) package contains such routines. In addition, a listing of a program that will perform the computations for the $1-M$ design is given as an appendix in Breslow and Day (1980).

7.2 Logistic Regression Analysis for the 1–1 Matched Study

The most frequently used matched design is one in which each case is matched to a single control. In this situation there are two subjects within each stratum. To simplify the notation, let x_{1k} denote the data vector for the case and x_{0k} the data vector for the control in the kth stratum. Using this notation, the conditional likelihood for the kth stratum is

$$l_k(\beta) = \frac{e^{\beta' x_{1k}}}{e^{\beta' x_{1k}} + e^{\beta' x_{0k}}} \tag{7.4}$$

Further simplification is obtained when we divide the numerator and denominator of equation (7.4) by $e^{\beta' x_{0k}}$ yielding

$$l_k(\beta) = \frac{e^{\beta'(x_{1k}-x_{0k})}}{1 + e^{\beta'(x_{1k}-x_{0k})}} \tag{7.5}$$

The expression on the right-hand side of equation (7.5) is identical to a logistic regression model with the constant term set equal to zero, $\beta_0 = 0$, and data vector equal to the value of the case minus the value of the control, $x_k^* = x_{1k} - x_{0k}$. This observation allows us to use standard logistic regression software to compute the conditional maximum likelihood estimates and obtain estimated standard errors of the estimated coefficients. To do this we define the sample size as the number of case-control pairs, use as covariates the differences x_k^*, set the values of the response variable equal to 1, $y_k = 1$, and exclude the constant term from the model. Thus, from a computational point of view, the 1–1 matched design presents no new challenges.

We have found that the process of creating the differences and setting the outcome equal to 1 can be confusing. It is important to distinguish between the logistic regression model being fit to the data and the computational manipulations used to apply standard logistic regression software. The process becomes less confusing when considering modeling strategies if we focus on terms in the logistic regression model first and then perform the computations needed to obtain the parameter estimates. A few examples should serve to illustrate this point.

First consider a dichotomous independent variable. It generates a single coefficient in the logit, irrespective of whether we enter the variable via a design variable or treat it as if it were continuous. If we code the variable as 0 or 1, then it follows that the difference variable, x^*, may take on one of three possible values, $(-1, 0, \text{or } 1)$. If we mistakenly thought of x^* as being the actual data we would have created two design variables. This should not have been done. Instead, the correct method is to create a new variable, x^*, which is computed as the difference between the two dichotomous variables in the pair and to then treat x^* as if it were continuous in the model.

As a second example, consider a variable such as race, coded at three levels. To correctly model this variable in the 1–1 matched design we would create, for each case and control in a pair, the values of the two design variables representing race. Then we would compute the difference between the case and control for each of these two design variables and treat each of these differences as if they were continuous. The same process is followed for any categorical scaled covariate. For example, suppose we wished to examine the scale in the logit of a continuous variable. The approach illustrated with the low birth weight data in Chapter 4 is to create design variables corresponding to the quartiles of the distribution and then plot their estimated coefficients. (Note: The quartiles come from the combined sample of 2K observations.) In the matched study we would do the same thing, with the one intermediate step of calculating the difference between the three design variables for case-control pairs. Because the computer does not recognize these differences in design variables as being from the same variable we will have to be sure that all three are included in any model we fit. One other point to keep in mind is that since the differences between variables used to form strata are zero for all strata, they will not enter any model in main effects form. However, we may include interaction terms between stratification variables and other covariates, as differences in these will likely not be zero.

In summary, the conceptual process for modeling matched data is identical to that already illustrated for unmatched data. If we develop our modeling strategies in the matched 1-1 design as if we had an unmatched design and then use the conditional likelihood, we will always be proceeding correctly.

7.3 An Example of the Use of the Logistic Regression Model in a 1-1 Matched Study

For illustrative purposes a 1-1 matched data set was created from the low birth weight data by randomly selecting for each woman who gave birth to a low birth weight baby, a mother of the same age who did not give birth to a low birth weight baby. For three of the young mothers (age less than 17) it was not possible to identify a match since there were no remaining mothers of normal weight babies of that age. The data set consists of 56 age matched case-control pairs. These data are listed in Appendix 3. With the exception of the number of first trimester visits, which has been excluded by us due to its lack of importance in the earlier analysis, the variables are the same as those in the low birth weight data set described in Table 4.1. In this example the number of prior preterm deliveries has been coded as a yes (1)-no (0) variable. Thus, at the initial stage of model building we have available the following variables: RACE, smoking status (SMOKE), presence of hypertension (HT), presence of uterine irritability (UI), presence of previous preterm delivery (PTD), and the weight of the mother at the last menstrual period (LWT). The variable AGE will be available when we evaluate interactions.

Before any model fitting was performed, we created difference variables. In all tables these variables will be denoted by the same words/letters as the original variable. While this could create some confusion, we feel it serves to emphasizes the fact that the logistic regression model is defined in terms of the original data rather than the difference variables. Obtaining the difference variables for RACE required that we apply the procedure described above. We created the two design variables for each case and control using white as the reference group. This is the same design variable coding as was used in Chapter 4. A difference variable was created for each design variable for RACE. All difference variables were entered in the model as if they were continuous.

Since there are only six variables, we begin model development with all variables in the model. The results of fitting this model are given in Table 7.1.

Table 7.1 Estimated Coefficients, Estimated Standard Errors, and Estimated Coefficient/Estimated Standard Error for the Model Containing All Variables.

Variable	Estimated Coefficient	Estimated Standard Error	Coeff./SE
RACE (1)	0.571	0.687	0.83
RACE (2)	−0.025	0.696	−0.04
SMOKE	1.401	0.624	2.25
HT	2.361	1.076	2.19
UI	1.402	0.693	2.02
PTD	1.808	0.780	2.32
LWT	−0.018	0.010	−1.80

Log-likelihood = −25.79

We see in Table 7.1 that neither design variable for RACE is significant, yet it may still be a confounder of the effects of the other variables in the model. To assess this, we refit the model without RACE and present the results in Table 7.2.

Table 7.2 Estimated Coefficients, Estimated Standard Errors, and Estimated Coefficient/Estimated Standard Error for the Model Excluding RACE.

Variable	Estimated Coefficient	Estimated Standard Error	Coeff./SE
SMOKE	1.479	0.558	2.65
HT	2.329	0.996	2.33
UI	1.345	0.691	1.95
PTD	1.670	0.741	2.25
LWT	−0.015	0.008	−1.86

Log-likelihood = −26.24

Comparing the estimated coefficients in Tables 7.1 and 7.2, we see that RACE seems to only confound the association for LWT, whose coefficient changes in value by over 20%. Because of this fairly substantial change, we proceed to the next step in which we identify the correct scale for LWT with RACE in the model.

To assess the scale of the variable LWT we create three design variables corresponding to the quartiles of the combined distribution of LWT ($n = 112$), using the first quartile as the reference group. This is the same method as was used in Chapter 4. As noted in Chapter 4, the nonparametric generalized additive modeling approach of Hastie and Tibshirani (1986, 1987) could have been used. However, in the absence of the specialized software needed to apply their method, the design variable approach should provide adequate guidance. The model is fit using the difference between the design variables for each case-control pair. *At this point we will drop reference to creation of design and difference variables and assume that when we discuss a variable in a model, it is implied that the design variable is created first and then differenced.*

The model was fit using all the variables shown in Table 4.1 except LWT, which has been replaced by the three design variables for quartiles. The estimated coefficients for the three design variables are given in Table 7.3.

Table 7.3   Results of the Quartile Analysis of LWT.

Quartile	1	2	3	4
Midpoint	93	113	128	176
Estimated Coefficient		−0.811	−0.430	−0.872
$\hat{\psi}$	1.0	0.44	0.65	0.42
95% CI		(0.1, 1.9)	(0.2, 2.2)	(0.1, 1.9)

The evidence for linearity in the logit for LWT is not apparent from the estimated coefficients shown in Table 7.3. If we plot the estimated coefficients versus the midpoint of the quartile using a value of zero for the first quartile, we see some linearity. However, the confidence interval estimates suggest that the effect of low weight is weak, and perhaps a dichotomization of LWT at the first quartile should also be tried. To assess this, a model treating LWT as a dichotomous variable coded 1 for the first quartile and zero elsewhere was fit. The deviances for these three codings of LWT are given in Table 7.4.

Based on the fact that the smallest deviance is obtained when LWT is treated as continuous and linear in the logit, we continue model development for LWT in this scale.

Table 7.4   Value of the Deviance for Three Different Codings of LWT.

Scale of LWT	Deviance
Linear	51.6
Quartiles	53.2
Dichotomous	53.4

Two additional models were fit to assess for evidence of nonlinearity in the logit for LWT. In the first model we added the square of LWT to the model. The addition of this variable did not reduce the deviance significantly. In the second model we added a variable which takes a more general look at the need for an additional term in the covariate. The use of this variable in matched designs was suggested by Pregibon (1984), and is in the same spirit as the transformation suggested by Box and Tidwell (1962) for normal theory models. The variable is of the form $x \ln(x)$. It is added to the model containing x to assess departure from linearity in x. A nonzero, significant, coefficient is evidence for nonlinearity. In our case the estimated coefficient for LWT $\times$ ln(LWT) was -0.06, with an estimated standard error of 0.06. Hence, we conclude that LWT can be adequately modeled as linear in the logit.

The next step in model development is to assess the possibility of interactions among the variables. A rather substantial number of potential interactions may be created from the variables in the model. In addition we would like to determine whether age interacts with any of the variables. Because of software limitations the design variables for RACE must be treated separately and interactions between RACE and the other variables will be discussed in terms of the individual design variables, RACE (1) and RACE (2). As is the case with the main effect variables, interaction variables are created first and then differenced. We do not form the interaction variables from the differences between main effect variables. In order to minimize the computational effort and maximize the number of interaction models examined, we used the method of performing a best subsets analysis described in Section 4.4.

We began the subset selection with a model containing the main effects, then added interactions to this model. This process identified only two interactions which were significant in any sense. These were the interactions of the design variable, RACE (1), for black versus white, with PTD and UI.

Since RACE is coded at three levels and forms two design variables, we fit a model containing the four interaction terms of RACE with PTD and UI. That is, we added the RACE (2) × PTD and RACE (2) × UI interactions to the model containing the RACE (1) × PTD and RACE (1) × UI interactions. The fit of this model provides us with another example of one of the limitations of many logistic regression software packages. As discussed in Section 4.5, these programs tend to produce results when they really should not. The estimated coefficient for the RACE (2) × UI interaction was 10.51 with a standard error of 25.79 which indicates the occurrence of a "zero cell." In this case, the problem may be traced to the fact that the differenced interaction variable takes on only the values zero and 1. Since this is a difference variable, values of −1, zero, and 1 are also required to obtain the estimate. The inclusion of the RACE (2) by PTD and UI interactions is not helpful. In addition, deletion of the main effect for RACE (2) from the model showed that it is not a confounder. Thus, we continued model development using only RACE (1) and its interactions. The implication of this decision is that RACE has been recoded to 1 = black and 0 = white or other.

Further examination of the contribution of RACE (1) and its interactions with PTD and UI showed that the crucial factor was the main effect for RACE (1). The inclusion of the interaction terms made the fitted model "unstable." We use the term unstable to mean a model whose estimated standard errors are much larger than the model without the variable. Additional analysis showed that the interactions were being determined by only three case-control pairs. The difference variables were zero except for these three pairs. Thus, we decided not to include these interactions in the model and proceeded to assess the fit of the model shown in Table 7.5.

Table 7.5 Estimated Coefficients, Estimated Standard Errors, and Estimated Coefficient/Estimated Standard Error for the Final Model.

Variable	Estimated Coefficient	Estimated Standard Error	Coeff./SE
RACE (1)	0.582	0.618	0.94
SMOKE	1.411	0.558	2.53
HT	2.351	1.041	2.26
UI	1.399	0.689	2.03
PTD	1.807	0.781	2.32
LWT	−0.018	0.009	−2.00

Log-likelihood = −25.80

7.4 Assessment of Fit in a 1–1 Matched Study

The approach to assessing the fit of a logistic regression model in the 1–1 matched design is identical to that described in Chapter 5 for unmatched designs. We begin by forming a measure of residual variation, and then using it to explore the sensitivity of the fit to individual case-control pairs. In the 1–1 matched study the likelihood function is defined in terms of the conditional probability of allocation of observed covariates to the case and control within each stratum. Using a value of $y = 1$ for the outcome in all strata (case-control pairs) corresponds to assigning a conditional probability of 1 to the observed allocation of covariate values to pairs. The fitted value is the estimate of this conditional probability under the assumption that the logistic regression model is correct. The number of covariate patterns will always be the number of pairs or strata. This implies that $m = 1$ for all patterns and measures that were based on m-asymptotics in the unmatched case cannot be used in 1–1 matched designs. For example, it is not possible to extend the Hosmer–Lemeshow chi-square goodness-of-fit statistic to the 1–1 matched study design

Moolgavkar, Lustbader, and Venzon (1985), and Pregibon (1984) have extended the ideas of Pregibon (1981) to matched studies. These authors show that, for 1–1 matched studies, the logistic regression diagnostics may be computed in the same manner as shown in Chapter 5 for unmatched studies. In particular, we may calculate leverage, standardized residuals, and the measures ΔX^2, ΔD and $\Delta \hat{\beta}$ using the formulae shown in equations (5.8)–(5.12), where

$x^* = x_{1k} - x_{0k}$ replaces x and we use the logistic model shown in equation (7.5).

The Pearson residual is

$$r = \frac{(y - \hat{\pi})}{[\hat{\pi}(1 - \hat{\pi})]^{1/2}}$$

and since $y = 1$ in the 1–1 matched design this simplifies to

$$r = \sqrt{\frac{(1 - \hat{\pi})}{\hat{\pi}}}$$

where $\hat{\pi}$ is the value of equation (7.5) using the estimated parameters. In this situation large residuals will only be possible when the fitted value, $\hat{\pi}$, is small. This was the same situation as in the unmatched case except that poor fit was also possible when $y = 0$ and the fitted value was large. Other than this, the observations on the expected behavior of the diagnostics as a function of the fitted values given in Table 5.3 also hold for the 1–1 matched study.

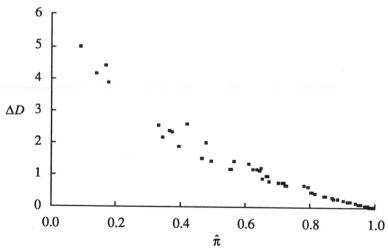

Figure 7.1   Plot of ΔD Versus $\hat{\pi}$.

To illustrate the use of the diagnostics, we apply them to assess the fit of the model whose estimated parameters are given in Table 7.5. The change in the deviance, ΔD, and the change in the Pearson chi-square, ΔX^2, due to deleting a particular pair, each show essentially the same thing so we only present plots for ΔD. Plots of ΔD and $\Delta \hat{\beta}$, versus the fitted values, $\hat{\pi}$, are shown in Figures 7.1–7.2.

In Figure 7.1 we see, as expected, that ΔD increases as $\hat{\pi}$ decreases. Four points have much larger values that the other points, and have corresponding values of $\hat{\pi}$ which are less than 0.2. By comparing Figure 7.2 to Figure 7.1 we can see the effect leverage has on the value of the influence diagnostic, $\Delta \hat{\beta}$. This follows from the fact, as shown in Chapter 5, that $\Delta \hat{\beta}$ is a function of both leverage and ΔX^2. Of the four points identified in Figure 7.1 only one has large (in fact, the largest) $\Delta \hat{\beta}$. A second point, not one of the four noted in Figure 7.1, has a value of $\Delta \hat{\beta} \approx 1$ with $\hat{\pi} \approx 0.4$ and although not shown, corresponds to the point with the maximum value of leverage.

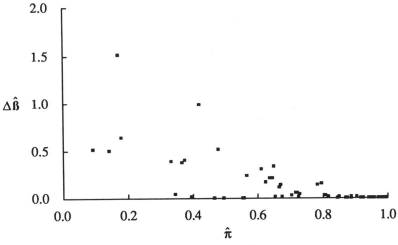

Figure 7.2   Plot of $\Delta \hat{\beta}$ Versus $\hat{\pi}$.

In Figure 7.3 we plot ΔX^2 versus $\hat{\pi}$ with the size of the plotting symbol proportional to $\Delta \hat{\beta}$.

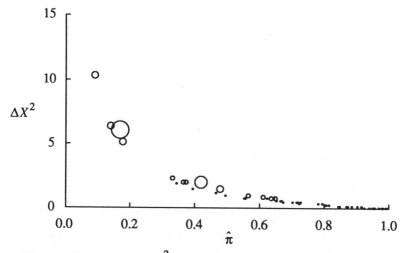

Figure 7.3   Plot of ΔX^2 Versus $\hat{\pi}$ with the Size of the Plotting Symbol
Proportional to $1.5 \times \Delta\hat{\beta}$.

As noted in Chapter 5 the values of $\Delta\hat{\beta}$ have been multiplied by 1.5 for
visual effect. In this figure the two points with the largest values of $\Delta\hat{\beta}$ stand
out. As noted, of the four points that are poorly fit (i.e., large ΔX^2 or ΔD),
only one has a large value of $\Delta\hat{\beta}$. As was the case in Chapter 5 this plot
presents a concise summary of all relevant diagnostic information.

These plots have identified a total of five pairs to be extreme (i.e., poorly
fit and/or influential). The values of the difference variables and diagnostics for
these five pairs are shown in Table 7.6.

Table 7.6   Data, Estimated Probability, and Value of the Five Diagnostic
Statistics $\Delta\hat{\beta}$, ΔX^2, ΔD, and Leverage (h) for Four Extreme Pairs ($P\#$).

P#	RACE (1)	SMOKE	HT	UI	PTD	LWT	$\hat{\pi}$	$\Delta\hat{\beta}$	ΔX^2	ΔD	h
9	0	−1	0	0	0	48	0.09	0.51	10.31	5.00	0.05
16	1	1	0	−1	−1	−49	0.42	0.98	2.04	2.57	0.33
18	0	−1	0	0	0	−12	0.15	0.50	6.34	4.17	0.07
27	0	1	−1	0	0	35	0.17	1.51	6.05	4.41	0.20
34	1	−1	0	0	0	38	0.18	0.63	5.14	3.86	0.11

Pair number 9 is poorly fit due to its very small fitted value, 0.09. Pairs 18 and 34 are also poorly fit, resulting from moderate leverages and moderate fitted values. None of these three pairs has a strong influence on the estimated coefficients so, without additional consideration of the biologic plausibility of their data, there is no reason to exclude them from the analysis. Pairs 16, and especially 27, have a substantial influence on the estimated coefficients. Pair 27 is also poorly fit. In order to demonstrate this we refit the model successively deleting these pairs. The results of this analysis are given in Table 7.7.

Table 7.7 Estimated Coefficients from All Pairs, Estimated Coefficients Obtained by Deleting Selected Pairs, and Values of the Deviance and Pearson Chi-Square Statistic.

Variable	All Data	Delete 16	% Change	Delete 27	% Change	Delete 16 and 27	% Change
RACE (1)	0.582	0.320	−45.0	0.803	37.9	0.536	−8.0
SMOKE	1.411	1.350	−4.3	1.316	−6.7	1.228	−13.0
HT	2.351	2.400	2.1	3.906	66.1	3.964	68.6
UI	1.399	1.742	24.5	1.639	17.2	2.005	43.3
PTD	1.807	2.146	18.8	2.067	14.4	2.443	35.2
LWT	−0.018	−0.015	−20.3	−0.027	47.3	−0.023	26.4
Goodness-of-fit							
D	51.59	49.18		46.6		44.21	
X^2	49.92	46.36		48.8		44.02	

Deletion of either or both pairs produces substantial changes in the estimated coefficients. We see that after deleting pair 27 the change in X^2 is less than expected; otherwise the results agree with those presented in Table 7.6. A final decision on whether to include or exclude any of these pairs will depend on an assessment of the biologic plausibility of the data. This decision should be made in consultation with subject matter experts.

7.5 An Example of the Use of the Logistic Regression Model in a 1–M Matched Study

The general approach to the analysis of the 1–M matched design and, for that matter, general matched or highly stratified designs is similar to that of the

1–1 matched design. Modeling problems are not conceptual but computational in nature. For the 1–1 matched design we were able to modify the data so that software for unconditional logistic regression could be used. However, for the 1–M design we need software that will maximize a more general conditional likelihood. The need for special software can be seen if we examine the contribution to the likelihood for an individual stratum. In this section, to keep notation simple and to a minimum, we will consider a design where $M = 3$. The extension of the methods to other matched designs is not difficult. We will let the value of the covariates for the case in stratum k be denoted by x_{1k} and the values for the three controls be denoted x_{2k}, x_{3k}, and x_{4k}. The contribution to the likelihood for this stratum is obtained by evaluating the expression shown in equation (7.3) and is

$$l_k(\beta) = \frac{e^{\beta' x_{1k}}}{e^{\beta' x_{1k}} + e^{\beta' x_{2k}} + e^{\beta' x_{3k}} + e^{\beta' x_{4k}}} \tag{7.6}$$

It is not possible to express the right-hand side of equation (7.6) in the form of a multigroup unconditional logistic regression model. (The multigroup, or polytomous logistic regression model, is presented in Chapter 8.) Hence, to perform an analysis of an 1–M matched design we must use software which obtains maximum likelihood estimators from a likelihood function whose component terms are like those in equation (7.6). The EGRET (1988) package provides such estimates. A listing of a program for obtaining conditional maximum likelihood estimates is given in the appendix of Breslow and Day (1980). Another approach is to employ a method for obtaining maximum likelihood estimates from a non-linear regression program. This technique is explained and illustrated in BMDP (1988) program BMDP3R. Given appropriate software, model development proceeds similarly to that described previously in this chapter for the 1–1 matched design.

For purposes of providing an example, a 1–3 matched design was created from the low birth weight data described in Section 1.5 and listed in Appendix 1. Mothers who gave birth to a low weight baby (cases) were randomly matched to three mothers of normal weight babies of the same age (controls). Twenty-nine strata of 1 case and 3 controls each were formed. Variables that were included for analysis in this example were the weight at the last menstrual period (LWT), smoking status (SMOKE), hypertension (HT), uterine irritability (UI), and presence or absence of a previous preterm delivery (PTD). These data are listed in Appendix 4. Since there are relatively few variables we begin model development with all variables in the model. The results of fitting a logistic regression model containing these variables are shown in Table 7.8.

Table 7.8   Estimated Coefficients, Estimated Standard Errors, and Estimated Coefficient/Estimated Standard Error for the Full Model.

Variable	Estimated Coefficient	Estimated Standard Error	Coeff./SE
SMOKE	0.554	0.488	1.13
HT	0.098	1.395	0.07
UI	0.525	0.549	0.96
PTD	1.532	0.638	2.40
LWT	−0.005	0.009	−0.60

Log-likelihood = −32.34

Before proceeding to assess which variables we should keep in the model we should investigate the scale of LWT in the logit. As in the 1−1 matched design the approach uses the estimated coefficients for the design variables for quartiles. In this case we form the quartiles from the combined distribution of 116 observations of LWT. Given the quartiles we form the values for the three design variables for each subject. These design variables are then used in the model in place of LWT. The estimated coefficients for the design variables obtained from a logistic regression model which also contains SMOKE, HT, UI, and PTD are given in Table 7.9.

Table 7.9   Results of the Quartile Analysis of LWT.

Quartile	1	2	3	4
Midpoint	98	116	130	195
Estimated Coefficient		−1.003	−0.033	−0.564
$\hat{\psi}$	1	0.37	0.97	0.57
95% CI		(0.1,1.6)	(0.2,3.9)	(0.1,2.3)

The estimated coefficients, odds ratios, and confidence intervals show that the logit is not linearly decreasing in LWT but women in the upper three quartiles are at slightly, but not significantly, less risk than those in the first quartile. On the basis of these observations a new model was fit containing the design variable for the first quartile, denoted LWD. The results of fitting this model are shown in Table 7.10

Table 7.10 Estimated Coefficients, Estimated Standard Errors, and Estimated Coefficient/Estimated Standard Error for the Model Treating LWT as Dichotomous.

Variable	Estimated Coefficient	Estimated Standard Error	Coeff./SE
SMOKE	0.555	0.483	1.15
HT	−0.026	1.305	−0.02
UI	0.500	0.541	0.92
PTD	1.525	0.636	2.40
LWD	0.519	0.525	0.99

Log-likelihood = −32.05

Comparing the results in Table 7.10 to those in Table 7.8 we see little change in any of the estimated coefficients. The Wald statistic for presence of hypertension, HT, is quite small and shows this variable may contribute little to the model. The results of fitting a model that excludes HT are shown in Table 7.11.

Table 7.11 Estimated Coefficients, Estimated Standard Errors, and Estimated Coefficient/Estimated Standard Error for the Reduced Model.

Variable	Estimated Coefficient	Estimated Standard Error	Coeff./SE
SMOKE	0.554	0.481	1.15
UI	0.500	0.541	0.92
PTD	1.526	0.635	2.40
LWD	0.521	0.515	1.01

Log-likelihood = −32.05

Comparison of the values of the estimated coefficients for the variables SMOKE, UI, PTD, and LWD between Tables 7.10 and 7.11 shows little evidence of confounding by HT. Exclusion of any variable remaining in the reduced model shown in Table 7.11 produces at least a 10% change in the estimated coefficient for one of the remaining variables, thus showing evidence of confounding. Hence, we proceed to assess for the need for interactions with the model shown in Table 7.11.

A total of 10 interaction variables were created. These included the interaction of AGE, the matching variable, with each of the four variables in the model. Each pairwise interaction among variables in the model was thought to have some biologic basis and thus was created as well. Among these 10 only the interactions of AGE with SMOKE and LWD were significant. The results of fitting a model containing the four main effects and these two interactions showed that only the AGE × LWD interaction was significant. Analysis of a model containing the four main effects and the AGE × LWD interaction showed that the estimated coefficients were sensitive to the inclusion or exclusion of one or two individuals. This provides an example of a model that may have an interaction component; but we cannot estimate it reliably with the available data. Hence, we proceed to assess the adequacy (i.e., goodness-of-fit) of the model containing only main effects as shown in Table 7.11.

7.6 Methods for Assessment of Fit in a 1–M Matched Study

The approach to assessment of fit in the 1–M matched study is similar to that used in the 1–1 matched study in that it is based on extensions of regression diagnostics for the unconditional logistic regression model. The mathematics required to develop these statistics is at a higher level than other sections of the book. Hence, less sophisticated mathematical readers may wish to skip this section and proceed to Section 7.7 where the use of the diagnostic statistics is explained and illustrated. These diagnostic statistics are derived for a general matched design by Moolgavkar, Lustbader, and Venzon (1985) and Pregibon (1984). These authors illustrate the use of the diagnostics only for the 1–1 matched design. We showed in Section 7.5 that the diagnostics for the 1–1 matched design may be computed using logistic regression software for the conditional model. Unfortunately, currently available software for logistic regression in the 1–M matched design does not compute these same diagnostic statistics. To obtain them we must extend the observations given in Section 5.3 on using a linear regression program to obtain logistic regression diagnostics. The method is, in principle, easy to apply; in practice, the computations necessary to calculate leverage values are tedious. Once the leverage values are obtained, the values of the other diagnostic statistics are calculated via simple transformations of available or easily computed quantities. To simplify the notation somewhat we present the methods for the case when $M = 3$; that is, $M + 1 = 4$.

The first step is to transform the observed values of the covariate vector by centering them about a weighted stratum-specific mean. That is, we compute for each stratum, k, and each subject within each stratum, j,

$$\tilde{\mathbf{x}}_{kj} = \mathbf{x}_{kj} - \sum_{l=1}^{4} \mathbf{x}_{kl}\hat{\xi}_{kl}$$

where

$$\hat{\xi}_{kj} = \frac{e^{\hat{\beta}'\mathbf{x}_{kj}}}{\sum\limits_{l=1}^{4} e^{\hat{\beta}'\mathbf{x}_{kl}}}$$

and note that $\sum_{j=1}^{4}\hat{\xi}_{kj} = 1$. Let $\tilde{\mathbf{X}}$ be the $n = 4K$ by p matrix whose rows are the values of $\tilde{\mathbf{x}}_{kj}$, $k = 1, 2, ..., K$ and $j = 1, 2, ..., 4$. Let $\mathbf{U}$ be an n by n diagonal matrix with general diagonal element $\hat{\xi}_{kj}$. It may be shown that the maximum likelihood estimate, $\hat{\beta}$ can be computed via the equation

$$\hat{\beta} = (\tilde{\mathbf{X}}'\mathbf{U}\tilde{\mathbf{X}})^{-1}\tilde{\mathbf{X}}'\mathbf{U}\mathbf{z}$$

where $\mathbf{z}$ is the vector of pseudovalues, $\mathbf{z} = \tilde{\mathbf{X}}\hat{\beta} + \mathbf{U}^{-1}(\mathbf{y} - \hat{\xi})$, $\mathbf{y}$ is the vector of values of the outcome variable ($y = 1$ for case and $y = 0$ for controls), and $\hat{\xi}$ is the vector whose components are $\hat{\xi}_{kj}$. Recall that $\hat{\xi}_{kj}$ is the estimated, under the assumption of a logistic regression model, conditional probability that subject j within stratum k is a case.

Thus, as was shown in Chapter 5 for the unconditional logistic regression model, we may recompute the maximum likelihood estimate for the conditional logistic regression model using a linear regression program allowing case weights. We use the vector $\tilde{\mathbf{x}}_{kj}$ as values of the independent variables,

$$z_{kj} = \tilde{\mathbf{x}}_{kj}' \, \hat{\boldsymbol{\beta}} + \frac{y_{kj} - \hat{\xi}_{kj}}{\hat{\xi}_{kj}}$$

as the values of the dependent variable, and case weight $\hat{\xi}_{kj}$, $k = 1, 2, ..., K, j = 1, 2, ..., 4$. It follows that the diagonal elements of the hat matrix computed by the linear regression are the leverage values we need,

$$h_{kj} = \hat{\xi}_{kj} \tilde{\mathbf{x}}_{kj}' (\tilde{\mathbf{X}}' \mathbf{U} \tilde{\mathbf{X}})^{-1} \tilde{\mathbf{x}}_{kj} \tag{7.7}$$

The standardized Pearson residual is

$$r_{kj} = \frac{(y_{kj} - \hat{\xi}_{kj})}{[\hat{\xi}_{kj} (1 - h_{kj})]^{1/2}}$$

which may be obtained from the residuals, weights, and leverage values from the linear regression as

$$r_{kj} = \frac{\sqrt{\hat{\xi}_{kj}} \, (z_{kj} - \hat{z}_{kj})}{\sqrt{1 - h_{kj}}}$$

In keeping with the diagnostics for the unmatched design we define the square of the standardized residual to be

$$\Delta X_{kj}^2 = r_{kj}^2 \tag{7.8}$$

and the influence diagnostic as

$$\Delta \hat{\boldsymbol{\beta}}_{kj} = \Delta X_{kj}^2 \frac{h_{kj}}{1 - h_{kj}} \tag{7.9}$$

The most informative way to view these diagnostics is via a plot of their values versus the fitted values, $\hat{\xi}_{kj}$. These plots are similar to those used in Chapter 5 to graphically assess the fit of the unconditional logistic regression model and those used in section 7.4 for the conditional logistical regression model in the 1–1 matched design. Examples of these plots are presented in the next section where we assess the fit of the model in Table 7.11.

Moolgavkar, Lustbader, and Venzon (1985) and Pregibon (1984) suggest that stratum specific totals of the two diagnostics, ΔX^2 and $\Delta \hat{\beta}$, be computed to assess what effect the data in an entire stratum has on the fit of the model. These statistics are computed as quadratic forms involving not only the leverage values for the subjects in the stratum but also those terms in the hat matrix which account for the correlation among the fitted values. An easily computed approximation to these statistics is obtained by ignoring the off diagonal elements in the hat matrix. The approximations are likely to be accurate enough for practical purposes. For the kth stratum these are are

$$\Delta X^2_k = r^2_k$$

$$= \sum_{j=1}^{4} r^2_{kj}$$

and

$$\Delta \hat{\beta}_k = \sum_{j=1}^{4} \Delta \hat{\beta}_{kj}$$

Strata with large values of these statistics would be judged to be poorly fit and/or have large influence respectively. A histogram or plot of their values versus stratum number will identify those strata with exceptionally large values. For these strata the individual contributions to these quantities should be examined carefully to determine whether cases and/or controls are the cause of the large values.

In identifying poorly fit or influential subjects deletion of the case in a stratum is tantamount to deletion of all subjects in the stratum. Without a case a stratum contributes no information to the likelihood function. If some but not all controls are deleted in a specific stratum then the stratum may still have enough information to contribute to the likelihood function. A final decision on exclusion or inclusion of cases (entire strata) or controls should be based on the biologic plausibility of the data.

7.7 An Example of Assessment of Fit in a 1–M Matched Study

Following the fit of the model whose estimated coefficients are presented in Table 7.11 we computed the values of the leverage show in equation (7.7) and the diagnostic statistics shown in equations (7.8) and (7.9). Plots of the diagnostic statistics versus the estimated probabilities, $\hat{\xi}_{kj}$, are shown in Figures 7.4 and 7.5, respectively.

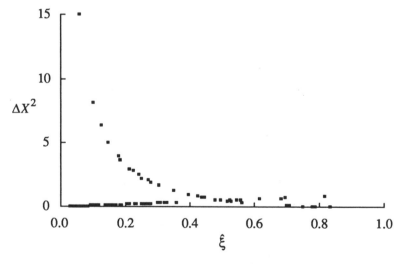

Figure 7.4 Plot of ΔX^2 Versus $\hat{\xi}$.

Examining the plot in Figure 7.4 we see a band of points at the bottom of the plot. These points correspond to all controls and those cases with large values of $\hat{\xi}$. We also see an arc of points sweeping up to the top left corner of the plot. The arc begins to be noticeable at about $\hat{\xi} = 0.4$. These points correspond to cases whose estimated values of $\hat{\xi}$ were small. The value of ΔX^2 for a poorly fit case is approximately $1/\hat{\xi}$. This follows from examining ΔX^2

when $y = 1$ and $\hat{\xi}$ approaches zero. Similarly, the value of ΔX^2 for a poorly fit control approaches $\hat{\xi}$. This is obtained by letting $y = 0$ and $\hat{\xi}$ approach 1. Thus, we see that poorly fit subjects, in the sense of having ΔX^2 exceed 1, are going to be cases and not controls. Of particular concern in Figure 7.4 is the subject (case) with a value of ΔX^2 approximately equal to 15.

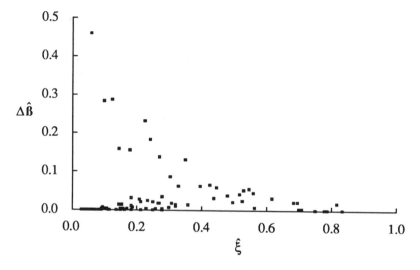

Figure 7.5 Plot of $\Delta\hat{\beta}$ Versus $\hat{\xi}$.

In Figure 7.5 the influence diagnostic, $\Delta\hat{\beta}$, is plotted versus $\hat{\xi}$ and the form of the plot is similar to Figure 7.4. The principal difference is the effect leverage may have on the value of the statistic. Moderate values of leverage and ΔX^2 can produce fairly large values of the influence diagnostic. However, in these data the dominant contributor to $\Delta\hat{\beta}$ is ΔX^2. The maximum value of $\Delta\hat{\beta}$ is approximately 0.5 and corresponds to the subject with the largest value of ΔX^2.

In Figure 7.6 we plot ΔX^2 versus $\hat{\xi}$ where the size of the plotted point is proportional to $1.5 \times \Delta\hat{\beta}$.

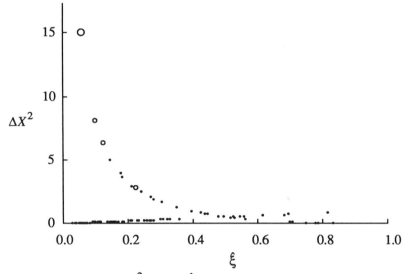

Figure 7.6 Plot of ΔX^2 Versus $\hat{\xi}$ Where the Size of the Plotted Symbol Is Proportional to $1.5 \times \Delta\hat{\beta}$.

Since none of the circles are large, this plot confirms that none of the subjects have exceptionally large values of $\Delta\hat{\beta}$. The subject with the largest circle is the one with the largest value of ΔX^2 and hence ΔX^2 is the major contributor to its value.

One subject (case) has been identified as being poorly fit and possibly influential. The data and diagnostic statistics for this case as well as the controls in the same strata are presented in Table 7.12.

In stratum 19 the case was poorly fit (small values of $\hat{\xi}$) largely due to the fact that the first control had a configuration of data and hence an estimated probability more like that of a case (large values of $\hat{\xi}$). The data for the case and the second and third control are identical. As a result, if the first control were removed, this stratum would contribute no information to the likelihood equation. Thus, to examine the effect of the first control on the model we delete the entire stratum. Note that if the data for the case and the other two controls had been different then we would have had the choice of deleting either the first control or the entire stratum. The results of fitting the model after deleting stratum 19 are presented in Table 7.13.

Table 7.12   Stratum Number, Data, Estimated Probability $\hat{\xi}$, Diagnostic Statistics $\Delta\hat{\beta}$, and ΔX^2, and Leverage (h) for the Most Extreme Subject and All Others in the Same Stratum.

Stratum	LOW	AGE	SMOKE	UI	PTD	LWD	$\hat{\xi}$	$\Delta\hat{\beta}$	ΔX^2	h
19	1	24	0	0	0	0	0.06	0.46	14.99	0.03
19	0	24	1	0	1	1	0.82	0.02	0.83	0.02
19	0	24	0	0	0	0	0.06	0.00	0.06	0.03
19	0	24	0	0	0	0	0.06	0.00	0.06	0.03

The results in Table 7.13 show the sensitivity of the estimated coefficients to the subjects deleted. In the model based on the data remaining after deleting stratum 19, the contribution of UI to the model is questionable. While not shown, its Wald statistic is 0.72. A reduced model which did not contain UI showed that it did not confound the associations of SMOKE, PTD, and LWD. At this point in the analysis we would have to make a choice. If we judge the data in stratum 19 to be plausible, albeit unusual, then we would use the model in Table 7.11. If we question the validity of the data in stratum 19 then we would likely use a model that includes only SMOKE, PTD, and LWD. A final decision on whether or not to exclude stratum 19 should be made in conjunction with subject matter experts.

Table 7.13   Estimated Coefficients from All Strata, Estimated Coefficients, and Percent Change from the All Strata Value When Stratum 19 Is Deleted.

Variable	All Strata	Delete Stratum 19	% Change
SMOKE	0.554	0.715	29.1
UI	0.500	0.401	−19.8
PTD	1.526	1.877	23.0
LWD	0.521	0.757	45.3
Log-Likelihood	−32.05	−28.95	

In summary the conceptual issues faced when using the logistic regression model in a matched design are identical to those encountered when using the model with an unmatched design. The problems are more computational in nature. Software for estimation and computation of diagnostic statistics for matched designs will likely be generally available in the near future. In the interim the methods presented in this chapter may be employed with currently available software.

7.8 The Benign Breast Disease Study

To provide a data set for exercises we present a subset of data from a large study on benign breast disease whose results have been published. The original data are from a hospital-based case-control study designed to examine the epidemiology of fibrocystic breast disease. Cases included women with a biopsy-confirmed diagnosis of fibrocystic breast disease identified through two hospitals in New Haven, Connecticut. Controls were selected from among patients admitted to the general surgery, orthopedic, or otolaryngologic services at the same two hospitals. Trained interviewers administered a standardized structured questionnaire to collect information from each subject [see Pastides et. al. (1983) and Pastides, et al. (1985)].

A code sheet for the data is given in Table 7.14 and they are listed in Appendix 5. Data are provided on 50 women who were diagnosed as having benign breast disease and 150 age matched controls, with three controls per case. Matching was based on the age of the subject at the time of interview.

Exercises

1. The data in Appendix 4 contains $1 - 3$ matched data constructed from the low birth weight data in Appendix 1. Using the case and the first control fit a logistic regression model for a $1 - 1$ matched design. (Note: It is also possible to use any one of the three controls. Designation of the first control was arbitrary.)

2. Using the case and controls 1 and 2 in Appendix 1, fit a logistic regression model for a $1 - 2$ matched design. (Note: It is also possible to use any two of the three controls. Designation of the first two controls was arbitrary.)

Table 7.14   Code Sheet for the Benign Breast Disease Data.

Variables	Name	Column Heading Appendix 5
1	Stratum (1–50)	STR
2	Observation within a Stratum (1 = Case, 2–4 = Control)	OBS
3	Age of the Subject at the Interview	AGMT
4	Final Diagnosis (1 = Case, 0 = Control)	FNDX
5	Highest Grade in School	HIGD
6	Degree (0 = None, 1 = High School, 2 = Jr. College, 3 = College, 4 = Masters, 5 = Doctoral)	DEG
7	Regular Medical Checkups (1= Yes, 2 = No)	CHK
8	Age at First Pregnancy	AGP1
9	Age at Menarche	AGMN
10	Number of Stillbirths, Miscarriages, etc.	NLV
11	Number of Live Births	LIV
12	Weight of the Subject	WT
13	Age at Last Menstrual Period	AGLP
14	Marital Status (1 = Married, 2 = Divorced, 3 = Separated, 4 = Widowed, 5 = Never Married)	MST

Missing values are denoted by "." in the listing of the data in Appendix 5.

3.   The data in Appendix 5 contain information from a 1 − 3 matched design studying benign breast disease. These data are described more fully in Section 7.8. Using the case and the first control fit a logistic regression model for a 1 − 1 matched design. (Note: It is also possible to use any one of the three controls. Designation of the first control was arbitrary.)

4.   Using the case and all controls in Appendix 5, fit a logistic regression model for a 1 − 3 matched design.

In each of the above problems the steps in fitting the model should include: (1) a complete univariate analysis, (2) an appropriate selection of variables for a multivariate model (this should include scale identification for continuous covariates and assessment of the need for interactions), (3) an assessment of fit of the multivariate model, (4) preparation and presentation of a table containing the results of the final model (this table should contain point and interval estimates for all relevant odds ratios), and (5) conclusions from the analysis.

CHAPTER 8

Special Topics

8.1 Polytomous Logistic Regression

8.1.1 Introduction to the Model and Estimation of the Parameters

Logistic regression is most frequently employed to model the relationship between a dichotomous (binary) outcome variable and a set of covariates, but with a few modifications it may also be used when the outcome variable is polytomous. We prefer to use the word binary in place of dichotomous when describing the number of outcome categories as it is more frequently used in the literature. This should not lead to any confusion. Some software packages such as SYSTAT offer the user the option of having an arbitrary number of outcome categories. The extension of the model and methods for a binary outcome variable to a polytomous outcome variable is easily illustrated when the outcome variable has three categories. Further generalization to an outcome variable with more than three categories is more of a notational problem than a conceptual one. Hence, we will consider only the situation when the outcome variable has three categories.

In developing models for a polytomous outcome variable we need to be aware of its measurement scale. Most applications, and hence the focus of the material in this section, involve a nominal scaled outcome variable. Methods are available for modeling an ordinal scale outcome variable but we will not present them. The interested reader should see McCullagh and Nelder (1983) for a concise discussion, with examples, of logistic regression as well as other types of models for ordinal scaled outcome variables.

Assume that the categories of the outcome variable, Y, are coded 0, 1, or 2. Recall that the logistic regression model for a binary outcome variable was parameterized in terms of the logit of $Y = 1$ versus $Y = 0$. In the three category model we have two logit functions: one for $Y = 1$ versus $Y = 0$, the other for $Y = 2$ versus $Y = 0$. In theory we could use any two of the pairwise logit comparisons of outcomes; but the obvious extension from the binary case is to use the logit of $Y = 2$ versus $Y = 0$ for the second function. Thus, the group coded $Y = 0$ will serve as the reference outcome value. The logit for comparing $Y = 2$ to $Y = 1$ may be obtained as the difference between the logit of $Y = 2$ versus $Y = 0$ and the logit of $Y = 1$ versus $Y = 0$.

Let $\mathbf{x}$ be the vector of covariates of length $p + 1$ with $x_0 = 1$ to account for the constant term. We will denote the two logit functions as

$$g_1(\mathbf{x}) = \ln\left[\frac{P(Y = 1 \mid \mathbf{x})}{P(Y = 0 \mid \mathbf{x})}\right]$$

$$= \beta_{10} + \beta_{11}x_1 + \beta_{12}x_2 + \cdots + \beta_{1p}x_p$$

$$= (1, \mathbf{x}')\,\boldsymbol{\beta}_1 \tag{8.1}$$

and

$$g_2(\mathbf{x}) = \ln\left[\frac{P(Y = 2 \mid \mathbf{x})}{P(Y = 0 \mid \mathbf{x})}\right]$$

$$= \beta_{20} + \beta_{21}x_1 + \beta_{22}x_2 + \cdots + \beta_{2p}x_p$$

$$= (1, \mathbf{x}')\,\boldsymbol{\beta}_2 \tag{8.2}$$

It follows that the three conditional probabilities of each outcome category given the covariate vector are

$$P(Y = 0 \mid \mathbf{x}) = \frac{1}{1 + e^{g_1(\mathbf{x})} + e^{g_2(\mathbf{x})}} \tag{8.3}$$

$$P(Y = 1 \mid \mathbf{x}) = \frac{e^{g_1(\mathbf{x})}}{1 + e^{g_1(\mathbf{x})} + e^{g_2(\mathbf{x})}} \tag{8.4}$$

and

$$P(Y = 2 \mid \mathbf{x}) = \frac{e^{g_2(\mathbf{x})}}{1 + e^{g_1(\mathbf{x})} + e^{g_2(\mathbf{x})}} \tag{8.5}$$

Following the convention for the binary model, we will let $\pi_j(\mathbf{x}) = P(Y = j \mid \mathbf{x})$ for $j = 0, 1, 2$ each of which is a function of the vector of $2(p + 1)$ parameters $\mathbf{\beta}' = (\mathbf{\beta}_1', \mathbf{\beta}_2')$.

A general expression for the conditional probability in the three category model is

$$P(Y = j \mid \mathbf{x}) = \frac{e^{g_j(\mathbf{x})}}{\displaystyle\sum_{k=0}^{2} e^{g_k(\mathbf{x})}}$$

where the vector $\mathbf{\beta}_0 = \mathbf{0}$ and hence $g_0(\mathbf{x}) = 0$.

To construct the likelihood function, it is convenient to formulate three binary variables coded as zero or 1 to indicate group membership of an observation. It should be noted that these variables are introduced only to clarify the likelihood function and are not constructed and used in the actual polytomous logistic regression analysis. The variables are coded as follows: if $Y = 0$ then $Y_0 = 1$, $Y_1 = 0$, and $Y_2 = 0$; if $Y = 1$ then $Y_0 = 0$, $Y_1 = 1$, and $Y_2 = 0$; and lastly if $Y = 2$ then $Y_0 = 0$, $Y_1 = 0$, and $Y_2 = 1$. We note that no matter what value Y takes on, $\Sigma Y_j = 1$. The conditional likelihood function for a sample of n independent observations is

$$l(\mathbf{\beta}) = \prod_{i=1}^{n} [\pi_0(\mathbf{x}_i)^{y_{0i}} \pi_1(\mathbf{x}_i)^{y_{1i}} \pi_2(\mathbf{x}_i)^{y_{2i}}]$$

Taking the log and using the fact that $\Sigma y_{ji} = 1$ for each i, the log-likelihood function is

$$L(\mathbf{\beta}) = \sum_{i=1}^{n} y_{1i} g_1(\mathbf{x}_i) + y_{2i} g_2(\mathbf{x}_i) - \ln(1 + e^{g_1(\mathbf{x}_i)} + e^{g_2(\mathbf{x}_i)}) \tag{8.6}$$

The likelihood equations are found by taking the first partial derivatives of $L(\beta)$ with respect to each of the $2(p + 1)$ unknown parameters. In order to simplify the notation somewhat we let $\pi_{ji} = \pi_j(x_i)$. The general form of these equations is as follows:

$$\frac{\partial L(\beta)}{\partial \beta_{jk}} = \sum_{i=1}^{n} x_{ki}(y_{ji} - \pi_{ji})$$

(8.7)

for $j = 1, 2$ and $k = 0, 1, 2, ..., p$; recall $x_{0i} = 1$ for each subject.

The maximum likelihood estimator, $\hat{\beta}$, is obtained by setting these equations equal to zero and solving for β. The solution requires the same type of iterative computation that is used in the binary outcome case.

The matrix of second partial derivatives is required to obtain the information matrix and asymptotic covariance matrix of the maximum likelihood estimator. The general form of the elements in the matrix of second partial derivatives is as follows:

$$\frac{\partial^2 L(\beta)}{\partial \beta_{jk} \partial \beta_{jk'}} = -\sum_{i=1}^{n} x_{k'i} x_{ki} \pi_{ji}(1 - \pi_{ji})$$

(8.8)

and

$$\frac{\partial^2 L(\beta)}{\partial \beta_{jk} \partial \beta_{j'k'}} = \sum_{i=1}^{n} x_{k'i} x_{ki} \pi_{ji} \pi_{j'i}$$

(8.9)

for j and $j' = 1, 2$ and k and $k' = 0, 1, 2, ..., p$. The information matrix, $I(\beta)$, is the $2(p + 1)$ by $2(p + 1)$ matrix whose elements are the negative of the expected values of the expressions given in equations (8.8) and (8.9). The asymptotic covariance matrix of the maximum likelihood estimator is the inverse of the information matrix, $\Sigma(\beta) = I(\beta)^{-1}$. The estimators of the information and covariance matrices are obtained by replacing the unknown parameters with the maximum likelihood estimators.

A more concise representation for the estimator of the information matrix may be obtained if we express it in a form similar to the binary outcome case. Let the matrix X be the n by $p + 1$ matrix containing the values of the covariates for each subject, let the matrix $\mathbf{V}_j$ be the n by n diagonal matrix with general element $\hat{\pi}_{ji}(1 - \hat{\pi}_{ji})$ for $j = 1, 2$ and $i = 1, 2, 3, ..., n$, and let $\mathbf{V}_3$ be the n by n diagonal matrix with general element $\hat{\pi}_{1i}\hat{\pi}_{2i}$. The estimator of the information matrix may be expressed as

$$\hat{\mathbf{I}}(\boldsymbol{\beta}) = \begin{bmatrix} \hat{\mathbf{I}}(\boldsymbol{\beta})_{11} & \hat{\mathbf{I}}(\boldsymbol{\beta})_{12} \\ \hat{\mathbf{I}}(\boldsymbol{\beta})_{21} & \hat{\mathbf{I}}(\boldsymbol{\beta})_{22} \end{bmatrix} \tag{8.10}$$

where

$$\hat{\mathbf{I}}(\boldsymbol{\beta})_{11} = (\mathbf{X}'\mathbf{V}_1\mathbf{X})$$

$$\hat{\mathbf{I}}(\boldsymbol{\beta})_{22} = (\mathbf{X}'\mathbf{V}_2\mathbf{X})$$

and

$$\hat{\mathbf{I}}(\boldsymbol{\beta})_{12} = \hat{\mathbf{I}}(\boldsymbol{\beta})_{21} = -(\mathbf{X}'\mathbf{V}_3\mathbf{X})$$

8.1.2 Interpreting and Assessing the Significance of the Estimated Coefficients

We begin by considering a model when we have a single dichotomous covariate which has been coded zero and 1. In the binary outcome model the estimated slope coefficient is identical to the log-odds ratio obtained from the 2 by 2 table cross-classifying the outcome and the covariate. The three outcome category model is parameterized in such a way that the two estimated coefficients, one from each logit function, are equal to the log-odds ratios from the pair of 2 by 2 tables obtained when cross-classifying the outcome by the covariate, and using the $y = 0$ group as the reference outcome value. Consider the following example.

A study was undertaken to assess factors associated with women's knowledge, attitude, and behavior toward mammography. These data have been made available to us by Dr. J. Zapka and Ms. D. Spotts of the University of Massachusetts, Division of Public Health. A code sheet for a few of the variables collected in a pilot study is given in Table 8.1.

As an example of a model with a dichotomous covariate, consider the cross-classification of mammography experience (ME) by HIST displayed in Table 8.2.

Table 8.1   Code Sheet for the Variables in the Mammography Experience Study.

Variable	Codes	Abbreviation in Appendix 6
Subject Number	1 – 412	OBS
Mammography Experience (Outcome Variable)	0 = Never 1 = Within the Past Year 2 = Over One Year Ago	ME
"You don't need a mammogram unless you develop symptoms"	1 = Strongly Agree 2 = Agree 3 = Disagree 4 = Strongly Disagree	SYMPT
Perceived Benefit of Mammography*	5 – 20	PB
Mother or Sister with a History of Breast Cancer	0 = No 1 = Yes	HIST
"Has anyone taught you how to examine your own breasts; that is BSE?"	0 = No 1 = Yes	BSE
"How likely is it that a mammogram could find a new case of breast cancer?"	1 = Not Likely 2 = Somewhat Likely 3 = Very Likely	DETC

*The variable PB is the sum of five scaled responses, each on a four point scale. A low value is indicative of a woman with strong agreement with the benefits of mammography.

In order to simplify the discussion of odds ratios in the polytomous outcome situation we need to generalize the notation used in the binary outcome case to include the outcomes being compared as well as the values of the

222   SPECIAL TOPICS

covariate. To accomplish this we assume that the outcome labeled with $Y = 0$ is the reference outcome. The subscript on the odds ratio indicates which outcome is being compared to the reference outcome. That is, the odds ratio of outcome j versus outcome 0 for covariate values of $x = a$ versus $x = b$ is

$$\psi_j(a,b) = \frac{P(Y = j \mid x = a)/P(Y = 0 \mid x = a)}{P(Y = j \mid x = b)/P(Y = 0 \mid x = b)}$$

Table 8.2 Cross-Classification of Mammography Experience (ME) by HIST and Estimated Odds Ratios Using Never as the Reference Outcome Value.

| | HIST | | | |
ME	0	1	TOTAL	$\hat{\psi}$
0	220	14	234	1.00
1	85	19	104	3.51
2	63	11	74	2.74
TOTAL	368	44	412	

In the special case when the covariate is binary, coded as zero and 1, we will let $\psi_j = \psi_j(1,0)$.

Using this notation, the two odds ratios calculated from Table 8.2 are

$$\hat{\psi}_1 = \frac{19 \times 220}{85 \times 14} = 3.51$$

and

$$\hat{\psi}_2 = \frac{11 \times 220}{63 \times 14} = 2.74$$

The results of fitting a three category logistic regression model to these data are presented in Table 8.3.

The values in the last column of Table 8.3, labeled $\hat{\psi}$, are obtained by exponentiation of the estimated slope coefficients. We note that they are identical to the values obtained from the 3 by 2 contingency table shown in Table 8.2. Confidence intervals are obtained in exactly the same manner as for the binary outcome model. A confidence interval for the coefficient is formed, and the endpoints are exponentiated to obtain the confidence interval for the odds ratio. For example, the 95% CI for the odds ratio of ME = 1 versus ME = 0 has endpoints

$$\exp(1.256 \pm 1.96 \times 0.375) = (1.69, 7.32)$$

Table 8.3   Results of Fitting the Logistic Regression Model to the Data in Table 8.2.

Logit	Variable	Estimated Coefficient	Estimated Standard Error	Coeff./SE	$\hat{\psi}$
1	HIST	1.256	0.375	3.35	3.51
	Constant	−0.951	0.128	−7.45	
2	HIST	1.009	0.428	2.36	2.74
	Constant	−1.251	0.143	−8.75	

Log-likelihood = −396.17

The log-odds for ME = 2 versus ME = 1 may be obtained from the difference between the two estimated slope coefficients in the logistic regression model. Illustrating this with the frequencies in Table 8.2 and the estimated coefficients from Table 8.3 we have

$$\ln\left(\frac{11 \times 85}{19 \times 63}\right) = 1.009 - 1.256 = -0.247$$

A preliminary indication of the importance of the variable may be obtained from the two Wald statistics; but as is the case with any multi-degree of freedom variable, we should use the likelihood ratio test to assess significance. For example, to test for the significance of the coefficients for HIST we compare the log-likelihood from the model containing HIST to the log-likelihood for the model containing only the two constant terms, one for each logit function. Under the null hypothesis that the coefficients are zero, minus twice the change in the log-likelihood will follow a chi-square distribution with 2 degrees of freedom. In the example the log-likelihood for the constant only model is −402.60. The value of the statistic is $-2 \times [-396.17 - (-402.60)] = 12.86$, which yields a p-value of 0.002. Thus, from a statistical point of view, the variable HIST is strongly associated with a woman's decision to have a mammogram.

In general, the likelihood ratio test for the significance of the coefficients for a variable will have degrees of freedom equal to the number of outcome categories minus one times the degrees of freedom for the variable in each logit. For example, if we have a four category outcome variable and a covariate that is modeled as continuous then the degrees of freedom will be $(4 - 1) \times 1 = 3$. If we have a categorical scaled covariate coded at five levels, then the variable will have four design variables within each logit and the degrees of freedom for the test will be $(4 - 1) \times (5 - 1) = 12$. This is easy to keep track of if we remember that we are modeling one logit for comparing the reference outcome category to each other outcome category.

For a polytomous covariate we expand the number of odds ratios to include comparisons of each level of the covariate to a reference level for each possible logit function. To illustrate this consider the variable DETC modeled via two design variables using the value of 1 (not likely) as the reference covariate value. The cross-classification of ME by DETC is given in Table 8.4.

Table 8.4 Cross-Classification of Mammography Experience (ME) by DETC.

ME	DETC 1	2	3	Total
0	13	77	144	234
1	1	12	91	104
2	4	16	54	74
Total	18	105	289	412

Using the value of ME = 0 as the reference outcome category and DETC = 1 as the reference covariate value, the four odds ratios are as follows:

$$\hat{\psi}_1(2,1) = \frac{12 \times 13}{77 \times 1} = 2.03$$

$$\hat{\psi}_1(3,1) = \frac{91 \times 13}{144 \times 1} = 8.22$$

$$\hat{\psi}_2(2,1) = \frac{16 \times 13}{77 \times 4} = 0.68$$

and

$$\hat{\psi}_2(3,1) = \frac{54 \times 13}{144 \times 4} = 1.22$$

The results of fitting the logistic regression model to these data are presented in Table 8.5.

Table 8.5 Results of Fitting the Logistic Regression Model to the Data in Table 8.4.

Logit	Variable	Estimated Coefficient	Estimated Standard Error	Coeff./SE	$\hat{\psi}$
1	DETC (1)	0.706	1.083	0.64	2.02
	DETC (2)	2.106	1.046	2.01	8.22
	Constant	−2.565	1.038	−2.47	
2	DETC (1)	−0.393	0.634	−0.62	0.68
	DETC (2)	0.198	0.594	0.33	1.22
	Constant	−1.179	0.572	−2.06	

Log-likelihood = −389.20

We see that exponentiation of the estimated logistic regression coefficients yields the odds ratios formed from 2 by 2 tables obtained from the main 3 by 3 contingency table. The odds ratios for logit 1 are obtained from the 2 by 3 table containing the rows corresponding to ME = 0 and ME = 1 and the 3 columns. The odds ratios for logit 2 are obtained from the 2 by 3 table containing the rows corresponding to ME = 0 and ME = 2 and the 3 columns.

To assess the significance of the variable DETC, we calculate minus twice the change in the log-likelihood relative to the constant only model. The value of the test statistic is $26.80 = -2 \times [-389.20 - (-402.60)]$ which, with 4 degrees of freedom, yields a p-value of less than 0.001. Thus, we would conclude that a woman's opinion on the ability of a mammogram to detect a new case of breast cancer is associated with her decision to have had a mammogram. Examining the estimated coefficients and their Wald statistics we see that the association is strongest when comparing the women who have had a mammogram within the last year to those who have never had one, and comparing the not likely to very likely response. All other estimated coefficients have nonsignificant Wald statistics. Confidence intervals for the odds ratios may be obtained in a manner similar to that shown for a dichotomous covariate.

Continuous scaled covariates which are modeled as linear in the logit will produce a single estimated coefficient in each logit function. This coefficient, when exponentiated, will give the estimated odds ratio for a change of one unit in the variable. Thus, remarks in Chapter 3 about knowing what a single unit is, and estimation of odds ratios for a biologically meaningful change apply directly to each logit function in the polytomous logistic regression model.

8.1.3 Model-Building Strategies for Polytomous Logistic Regression

In principle, the strategies and methods for multivariate modeling with a polytomous outcome variable are identical to those for the binary outcome variable discussed in Chapter 4. The theory for stepwise selection of variables has been worked out and is available in some site-specific programs, such as the one documented in Hosmer, et. al (1978). The method is not currently available in the widely distributed statistical software packages. To illustrate modeling and interpretation of the results, we proceed with an analysis of the data from the mammography study.

The analysis of the data from the mammography study is relatively simple as there are only five independent variables and 412 subjects. We do have a few decisions to make regarding how some of the variables are going to be entered into the model. In particular, the variable SYMPT is coded at four levels on an ordinal scale. Traditionally, variables of this type have either been analyzed as if they were continuous or categorical. We will begin the model-building process with SYMPT coded into three design variables, using the "strongly agree" response as the reference value. The variable DETC is coded at three levels and is ordinal scaled. It will also be entered into the model using two design variables with the "not likely" response as the reference value. The rationale for coding these ordinal scaled variables into design variables rather than treating them as if they were continuous is that the coefficients for the design variables may be plotted to assess the functional form of the two logits over the categories. Initially, we will treat the variable PB as if it were continuous and

linear in the logits. The results of fitting the full multivariate model are given
in Table 8.6.

Examination of the Wald statistics in Table 8.4 suggests that, with the
possible exception of the variable DETC, each of the variables may contribute to
the model. For the moment we will keep variables in the model while we
examine the scale of the variable SYMPT.

Table 8.6 Estimated Logistic Regression Coefficients, Estimated Standard
Errors, and Estimated Coefficient/Estimated Standard Error for the Full
Multivariate Model.

Logit	Variable	Estimated Coefficient	Estimated Standard Error	Coeff./SE
1	SYMPT (1)	0.110	0.923	0.12
	SYMPT (2)	1.925	0.778	2.48
	SYMPT (3)	2.457	0.775	3.17
	PB	−0.219	0.076	−2.91
	HIST	1.366	0.438	3.12
	BSE	1.292	0.530	2.44
	DETC (1)	0.017	1.162	0.01
	DETC (2)	0.904	1.127	0.80
	Constant	−2.999	1.539	−1.95
2	SYMPT (1)	−0.290	0.644	−0.45
	SYMPT (2)	0.817	0.540	1.51
	SYMPT (3)	1.130	0.548	2.06
	PB	−0.148	0.076	−1.94
	HIST	1.065	0.459	2.32
	BSE	1.052	0.515	2.04
	DETC (1)	−0.924	0.714	−1.29
	DETC (2)	−0.691	0.687	−1.01
	Constant	−0.986	1.111	−0.89

Log-likelihood $= -346.95$

The two estimated coefficients for the design variable SYMP(1), which
estimates the log odds for the response agree versus the reference value of
strongly agree, suggest that these two categories are similar and their Wald
statistics are significant. The sign and magnitude of the estimated coefficients

for the design variables SYMPT(2) and SYMPT(3) suggest that the responses disagree and strongly disagree differ from strongly agree and are also similar within each of the two logit functions. This type of pattern in the estimated coefficients does not support treating the variable as continuous and linear in the logit; rather it suggests that we should dichotomize SYMPT into two levels, coded $0 =$ strongly agree or agree and $1 =$ disagree or strongly disagree, and refit the model. This new dichotomous variable is labeled SYMPD in the output. The log-likelihood for the model containing SYMPD in place of SYMPT is decreased to -348.75. We proceed to the next step, assessment of the scale of PB, using SYMPTD.

To assess the scale of the variable PB we use the method of creating design variables used previously. The values of PB are integers and range from 5 to 17, with relatively few exceeding 10. PB was broken into six categories corresponding to the values of 5, 6, ..., 9, and ≥ 10. Five design variables were formed using PB $= 5$ as the reference value. A plot of the estimated logistic regression coefficients for the five design variables from the two logit functions is given in Figure 8.1. It should be noted that the model contained all the other variables.

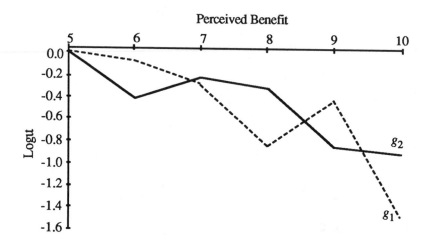

Figure 8.1 Plot of the Estimated Logistic Regression Coefficients for the Five Design Variables Created from PB for the Two Logit Functions Denoted g_1 and g_2 in the Figure.

The polygons for the two logit functions in Figure 8.1 show no strong evidence against the logits being linear in PB. Hence, we choose to include the variable PB in the model as continuous (not recoded) and linear in each of the two logit functions.

It remains to assess the contribution to the model and scale of the variable DETC. The results of fitting the model using SYMPD and treating PB as continuous are given in Table 8.7.

Table 8.7 Estimated Logistic Regression Coefficients, Estimated Standard Errors, and Estimated Coefficient/Estimated Standard Error for the Reduced Model Treating SYMPD.

Logit	Variable	Estimated Coefficient	Estimated Standard Error	Coeff./SE
1	SYMPD	2.095	0.457	4.58
	PB	−0.251	0.073	−3.44
	HIST	1.293	0.434	2.98
	BSE	1.244	0.526	2.36
	DETC (1)	0.090	1.161	0.08
	DETC (2)	0.973	1.126	0.86
	Constant	−2.704	1.434	−1.88
2	SYMPD	1.121	0.357	3.14
	PB	−0.168	0.074	−2.27
	HIST	1.014	0.454	2.23
	BSE	1.029	0.514	2.00
	DETC (1)	−0.902	0.715	−1.26
	DETC (2)	−0.669	0.688	−0.97
	Constant	−0.999	1.072	−0.93

Log-likelihood = −348.75

In Table 8.7 we see that none of the Wald statistics for the design variables for DETC are significant; but, taken in total, the variable could still be significant and also a confounder for some of the other variables. To assess its significance and role as a confounder we fit the model excluding DETC. The log-likelihood for this model is −353.02. The likelihood ratio test statistic for the significance of DETC is 8.54 = −2[−348.75 − (−353.02)] which, with 4 degrees of freedom, yields a p-value of 0.074. Among the variables remaining in the model the greatest change in an estimated coefficient was for SYMPD in logit 1. The change was about 8% of the value of the estimated coefficient, indicating that DETCT plays some role as a confounder of the association of SYMPD with the outcome variable. Because of its role as a modest confounder and its near statistical significance, we decide to keep DETC in the model. In addition, we choose to model via design variables as there is no pattern in the estimated coefficients for the design variables that suggests that the variable should be treated as continuous and linear in the logit.

As was the case for a binary outcome model the next step in model development is to assess the need to include interaction terms in the model. In the mammography experience study each pair of variables creates a biologically plausible interaction. None of the 10 possible interactions contributed significantly to the model. Thus, we take as our final model the one displayed in Table 8.7.

Begg and Gray (1984) have proposed a method for approximating the fit of a polytomous logistic regression via fitting individual binary logistic regression models. For example, in a three group problem we would fit a model for $Y = 1$ versus $Y = 0$ using a standard logistic regression method for a binary outcome variable and fit separately a model for $Y = 2$ versus $Y = 0$. The coefficients for the polytomous model are obtained from the two separately fit logistic models. Begg and Gray show that the estimates of the logistic regression coefficients obtained in this manner are consistent, and under many circumstances the loss in efficiency is not too great. It has been our experience that the coefficients obtained from separately fit logistic models will be close to those from the polytomous fit. Thus, the individualized fitting approach can be useful for variable selection; but we would urge some caution in drawing definitive inferences using these estimates and their associated estimated standard errors. If at all possible, final inferences should be based on estimated coefficients and estimated standard errors from fitting the polytomous logistic regression model.

To illustrate the Begg and Gray method of fitting individual logistic regressions we used it to refit the model shown in Table 8.7. The results of this fit along with the maximum likelihood fit are given in Table 8.8. In Table 8.8

the columns labeled ILR give the estimated coefficients from the individualized logistic regressions and the maximum likelihood estimates are given in the columns labeled MLE.

Table 8.8 Comparison of the Maximum Likelihood Estimates, MLE, and the Estimates from Individual Logistic Regression Fits, ILR.

		Estimated Coefficient		Estimated Standard Error	
Logit	Variable	MLE	ILR	MLE	ILR
1	SYMPD	2.095	2.093	0.457	0.465
	PB	−0.251	−0.242	0.073	0.074
	HIST	1.293	1.388	0.434	0.470
	BSE	1.244	1.357	0.526	0.535
	DETC (1)	0.090	0.269	1.161	1.159
	DETC (2)	0.973	1.096	1.126	1.119
	Constant	−2.704	−3.012	1.434	1.427
2	SYMPD	1.121	1.137	0.357	0.357
	PB	−0.168	−0.164	0.074	0.074
	HIST	1.014	1.037	0.454	0.462
	BSE	1.029	1.013	0.514	0.515
	DETC (1)	−0.902	−0.749	0.715	0.717
	DETC (2)	−0.669	−0.545	0.688	0.686
	Constant	−0.999	−1.158	1.072	1.075

Comparing the pairs of columns in Table 8.8, one set for estimated coefficients and the other for estimated standard errors, we see that the method of individual logistic regressions proposed by Begg and Grey provides a good approximation to both the estimated coefficients and estimated standard errors. Thus, in the absence of software to perform a polytomous outcome analysis, we could use the results of individual logistic regressions, realizing of course that the resulting estimates are approximations to the maximum likelihood estimates.

One problem which we were not faced with in the binary outcome case but which can be an issue in a polytomous logistic regression model is the situation where the covariate is significant for some but not all logit functions. If we model using the principle that we would like to minimize the number of parameters, then we should force the coefficients to be zero in some logit functions and estimate their values for the other logit functions. This strategy is not possible with currently available polytomous logistic regression software, but can be accommodated using the individualized logistic regression approach. As in all modeling situations biologic considerations should play an important role in variable selection.

Finally, if the analysis is performed via individual logistic regressions, we may employ currently available software and use the variable selection strategies described in Chapter 4 for each logit function. As noted above, this approach may result in different variables for different logit functions and this would have to be considered in formulating a final model.

8.1.4 Assessment of Fit and Diagnostics for Polytomous Logistic Regression Model

As is the case with any fitted model, before using it to make inferences we should assess its overall fit and examine the contribution to the fit of each subject. In polytomous logistic regression the multiple outcome categories make this a more difficult problem than was the case with a model for a binary outcome variable. When we model a binary outcome variable we have a single fitted value, the estimated logistic probability of the outcome being present, $P(Y = 1 \,|\, x)$. When the outcome variable has three categories we have two estimated logistic probabilities, the estimated probabilities of categories 1 and 2, $P(Y = 1 \,|\, x)$ and $P(Y = 2 \,|\, x)$. Lesaffre (1986) has proposed extensions of tests for goodness-of-fit and logistic regression diagnostics to the polytomous logistic regression model. However, these methods are not easily calculated using available software. Thus, we recommend assessing fit and calculating logistic regression diagnostics using the individual logistic regressions approach of Begg and Grey.

For an outcome variable with three categories we would assess the fit of the two logistic regression models and then integrate the results, usually descriptively, to make a statement about the fit of the polytomous logistic regression model. The procedure for assessing the fit of each individual logistic regression model would use the methods described in Chapter 5. Integration of the results requires thoughtful consideration of the effects of influential and poorly fit covariate patterns on each model. In particular, covariate patterns that are influential for only one logit should be examined closely with due consideration to biologic issues before they are excluded from analyses. While this process requires more computation than for a single logistic regression model for a binary outcome variable, there is nothing new conceptually.

We illustrate the methods by considering assessment of fit of the polytomous logistic regression model shown in Table 8.8 for the mammography experience study. The approach is to evaluate the fit of the individual logistic regressions shown in Table 8.8 and then integrate these results to assess the fit of the polytomous model. Summary goodness-of-fit statistics are presented in Table 8.9 for each of the individual logistic regression models. Logit model 1 refers to the logistic regression comparing the women who had a mammogram within a year of the interview (ME = 1) to the never had a mammogram group (ME = 0) and logit model 2 compares the women who had a mammogram over 1 year prior to the interview (ME = 2) to the never had a mammogram group (ME = 0). These statistics are calculated using the observed covariate patterns generated by the variables in the model. For logit model 1 there were $J = 81$ patterns and for logit model 2 there were $J = 83$ patterns.

Table 8.9 Summary Goodness-of-Fit Statistics for the Individual Logistic Regressions.

Logit	HL (C)	Deviance (D)	Pearson (X^2)
Model 1	6.91	51.91	118.00
df (p–value)	8 (0.55)	74	74
Model 2	12.99	70.90	63.75
df (p–value)	8 (0.11)	76	76

The Hosmer–Lemeshow, HL (C), statistics have values of 6.91 (df = 8, p = 0.55) and 12.99 (df = 8, p = 0.11) and indicate that there is satisfactory overall fit for both models. The deviances, D, also indicate that the models seem to fit as they are less than their respective degrees of freedom, for example, 51.91 < 74 and 70.90 < 76. We do not provide p-values for the deviance as the number of subjects per covariate pattern is not large enough in either model for m-asymptotics to hold. The Pearson chi-square statistic for logit model 2 is also less than its degrees of freedom, 63.75 < 76. However, for logit model 1 the value is 118.0 which is considerable larger than its degrees of freedom. The discrepancy between values of the deviance and the Pearson chi-square statistic for logit model 1 is certainly unusual and indicates that we should examine the diagnostic statistics to determine the reason for such a large difference.

The leverage, h, and diagnostic statistics $\Delta\hat{\beta}$, ΔX^2, and ΔD defined in equations (5.8) and (5.10)–(5.12) were calculated for each covariate pattern for each of the two individual logistic regression models. Plots similar to those shown in Chapter 5 identified several patterns with large values on one or more statistics. Information for these patterns is summarized in Table 8.10. The quantity P# is an arbitrary designation for covariate pattern. Its value depends on the order in which the covariate patterns are formed. Pattern numbers are provided to facilitate discussion of the values of the diagnostic statistics.

Table 8.10 Data, Observed Outcome (y), Number (m), Estimated Logistic Probability $\hat{\pi}$, and the Value of the Three Diagnostic Statistics $\Delta\hat{\beta}$, ΔX^2, ΔD, and Leverage (h), for Influential or Poorly Fit Covariate Patterns from Each Individual Logistic Regression Model.

Logit Model	P#	SYMPD	PB	HIST	BSE	DETC	y	m	$\hat{\pi}$	$\Delta\hat{\beta}$	ΔX^2	ΔD	h
1	5	0	6	0	0	2	1	2	0.015	0.55	32.67	5.77	0.02
	28	0	10	0	1	1	1	1	0.017	1.40	59.98	8.37	0.02
	68	0	9	0	1	3	11	18	0.345	1.72	7.01	6.59	0.20
2	43	1	5	0	0	3	1	1	0.169	0.29	5.17	3.75	0.05
	70	1	10	0	0	3	2	6	0.099	0.82	4.36	2.89	0.16
	71	1	10	0	0	1	1	1	0.159	0.47	5.68	3.97	0.08
	72	1	10	0	1	3	2	19	0.233	0.97	2.43	2.89	0.29

Examining the diagnostic statistics for logit model 1 we see why the value of the Pearson chi-square was so large relative to the deviance. Patterns 5 and 28 have very small estimated logistic probabilities and observed probabilities, y/m, of 0.5 and 1.0, respectively. These differences generate extremely large Pearson residuals. The deviance residuals, while not as large, are also considered significant. The extremely large value of ΔX^2 is the primary reason that $\Delta\hat{\beta}$ is so large for covariate pattern 28. This same situation was encountered when assessing the fit of the logistic regression model for the low birth weight data in Chapter 5: namely covariate patterns with a small estimated probability and an outcome that was contrary to the model. Based on the biologic plausibility of the observed covariates we have little reason to exclude these patterns from the data.

Covariate pattern 68 for logit model 1 presents a new challenge in assessing the fit of a model. The responses to the variables SYMPD, HIST, BSE, and DETC in this pattern are what we might call a "modal" response. The observed pattern in these variables represents a woman who disagrees (SYMPD = 1) with the statement, "You don't need a mammogram unless you develop symptoms," has no family history of disease (HIST = 0), has been taught breast self-examination (BSE = 1), and believes that it is very likely for mammography to detect a new case of breast cancer (DETC = 3). In fact, 149 of the 338 subjects used in fitting logit model 1 had this particular response to these variables. The single remaining variable to differentiate outcome among these subjects is the scaled variable PB and covariate pattern 68 corresponds to PB = 9. For this covariate pattern the value of $\Delta\hat{\beta}$ is 1.72 which is quite large. This agrees with the expectations set out in Table 5.3 of a pattern with moderate leverage and change in Pearson chi-square. We would not want to discard data on 18 subjects representing a fairly common response pattern without first trying to improve the model. On the other hand, we have little other additional information in the covariates. Addition of all interaction terms of PB with the other main effects did not change the deviance significantly for logit model 1. The deviance for the interactions model is 48.99 versus 51.91 for the main effects model in Table 8.8, yielding a likelihood ratio test statistic of 2.92 with df = 5 and $p = 0.71$. Including the interactions increased the sensitivity of the model to covariate pattern 68 as the value of $\Delta\hat{\beta}$ increased to 1.96. There are seven other covariate patterns with the same "modal" response as pattern 68. The logistic regression model fit each of these other patterns adequately. For these patterns the value of PB ranged from 5 to 12 so PB = 9 for pattern 68 would not be considered to be an extreme response.

At this point we have to decide what to do with pattern 68. Deletion of this pattern from the data results in a decrease in the deviance of 6.72 which with 1 degree of freedom yields a p-value of 0.01. Deletion of a covariate pattern is analytically equivalent to the creation of a design variable that takes the value of 1 for the covariate pattern and zero otherwise, and adding this design variable to the model. Thus, the change in deviance may be used to test the hypothesis that the coefficient for this design variable is zero. This reduced model still did not fit patterns 5 and 28 well but the Pearson chi-square did decrease to 106.7. Since we still have to assess the fit of the logit model 2, we will defer a final decision on covariate pattern 68 until that is completed.

Before considering the diagnostic statistics for logit model 2, we point out the fact that covariate pattern 68 in logit model 1 provides an excellent example of why diagnostics should be calculated by covariate patterns formed from the main effects in a model rather than for individual cases. Had we considered the 18 subjects with covariate pattern 68 individually an entirely different picture would emerge. First, the leverage for each of the 18 subjects would be $0.0109 = 0.197/18$. For the 11 subjects with the response present the diagnostic statistics would have had values of $\Delta X^2 = 1.92$ and $\Delta \hat{\beta} = 0.021$, which would indicate some lack-of-fit but little influence on the estimated coefficients. The 9 subjects with the response absent would have had $\Delta X^2 = 0.53$ and $\Delta \hat{\beta} = 0.006$, which would support an adequate model. Thus, had we considered the data on an individual basis, we would have missed an important source of lack-of-fit and influence on the estimated coefficients.

Examining the diagnostic statistics in Table 8.10 for logit model 2 we see that patterns 43, 70, and 71 are poorly fit. It should be noted that computations for the two individual logistic regressions were performed on separate data sets, thus the pattern numbers for the two models do not refer to the same covariate patterns. In the case of patterns 43 and 71 the estimated logistic probabilities are small but not less than 0.1. The values of the covariates are reasonable and, as was the case with patterns 5 and 28 for logit model 1, the observed outcome was contrary to the model.

Pattern 72 is the one with the largest value of $\Delta \hat{\beta}$. This pattern also represents the "modal" response described above in our discussion of the fit of logit model 1. In the case of logit model 2 a total of 115 of the 308 subjects used in the analysis had the "modal" response pattern. For this pattern the value of PB is 10. Patterns 70 and 71 also have PB = 10. In Table 8.10 we see that nearly all the patterns with extreme values of the diagnostic statistics have a value of PB of either 9 or 10. It appears that the response of subjects with PB in this range is more variable than the logistic model is able to account for. As was the case with logit model 1 the addition of interaction terms to logit model 2 did not improve the fit.

At this point in the analysis we have few options with the available data. No alternative model was able to improve on the model shown in Table 8.7. To explore the effect the two influential covariate patterns have on the model we eliminate all subjects with data corresponding to covariate patterns 68 and 72 in Table 8.10 and refit the model. Results are presented in Table 8.11.

Table 8.11 Estimated Logistic Regression Coefficients, Estimated Standard Errors, and Estimated Coefficient/Estimated Standard Error After Deleting All Subjects with Covariate Patterns 68 and 72.

Logit	Variable	Estimated Coefficient	Estimated Standard Error	Coeff./SE
1	SYMPD	2.120	0.463	5.58
	PB	−0.223	0.086	−2.59
	HIST	1.284	0.436	2.94
	BSE	1.273	0.532	2.39
	DETC (1)	0.126	1.162	0.11
	DETC (2)	1.051	1.135	0.92
	Constant	−2.998	1.528	−1.96
2	SYMPD	1.276	0.361	3.54
	PB	−0.080	0.079	−1.01
	HIST	0.897	0.456	1.97
	BSE	1.191	0.517	2.30
	DETC (1)	−0.812	0.711	−1.14
	DETC (2)	−0.365	0.692	−0.53
	Constant	−2.029	1.147	−1.77

Log-likelihood = −316.69

Comparing the estimated coefficients in Tables 8.7 and 8.11 we see that the order of magnitude of the coefficients for the most part has not changed. However, there are changes in some coefficients substantial enough to change odds ratio estimates.

The problems encountered in fitting the model to patterns with PB in the middle of its range point out one of the dangers in using summary indices. The variable perceived benefit, PB, was created from five other variables. Considerable variability, and hence, information, about the responses of individual subjects may be lost. In the current example a value of PB = 9 or 10 could have been obtained from many possible combinations of responses to its five component variables. In situations when the original data are available a prudent strategy would be to remove the summary variable and consider as covariates the individual components.

8.2 Logistic Regression Analysis of Survival Data

In Chapter 6 we developed the likelihood function for use of the logistic regression model in the cohort study. One fundamental characteristic of this study design is that the values of the covariates are theoretically known to the investigator prior to determination of the outcome variable of interest. An example of a cohort study is a clinical trial to study the effectiveness of two therapies to treat cancer. In this study we may randomize subjects to one of the two treatments and determine the values of those covariates that are known, or thought, to influence the effectiveness of treatment. In studies of this type the outcome variable of interest may not simply be the occurrence or nonoccurrence of an event, for example, "death." Instead, interest may focus on the length of time to the event, termed the "survival time." The terms "follow-up time" and "failure time" have also been used, among others, to label this outcome. We will use survival time in this section and call the occurrence of the event "death." Examples of studies where the survival time is the variable of interest are not limited to clinical trials. The goal of the analysis in these studies is to describe the association between "treatment" and the length of survival time controlling for patient-specific covariates. Cox (1972) proposed the proportional hazards model for performing a regression analysis of survival time on set of covariates. This model has become the standard method in many fields for analyses of this type. There is an enormous literature dealing with the subject and a number of methods textbooks [e.g., Cox and Oakes (1984), Kalbfleisch and Prentice (1980), and Lee (1980)].

On occasion the occurrence or nonoccurrence of the event has been modeled in studies where survival time is the primary outcome variable using logistic regression. The reasons some investigators choose to use logistic regression have undoubtedly been practical ones, such as not having the software necessary to use a proportional hazards model analysis, and/or not feeling as comfortable with the concepts behind a proportional hazards model analysis as with logistic regression analysis. It is not the goal of this section to provide an introduction to the analysis of survival time. Instead, we indicate how and when logistic regression may be used when survival time is really the outcome of interest.

The use of logistic regression with survival data has been studied recently by Efron (1988), Abbott (1985), Brenn and Arnesen (1985), Green and Symmons (1983), and Hauck (1985). The rationale for using the logistic regression model comes from another model proposed by Cox (1972), called the discrete linear logistic, to handle the case when the survival time is either discrete or known only for a few discrete time intervals, for example, 6-month intervals in a study of 2-year survival. In most health sciences applications time of an event is not known exactly but is known only to have occurred in an interval. We will use the term interval to indicate our observation of survival time. The logistic model assumes that the logit of the probability of death in an interval, conditional that death has not occurred prior to that interval, is a linear function of the covariates and a constant term specific to the interval. The proportional hazards model assumes that $\ln[-\ln(p)]$, known as the complimentary log-log transformation, is the linearizing transformation. When the conditional probabilities are small for each interval (<0.1), both the discrete proportional hazards model and the logit model tend to yield similar results. Wu and Ware (1979) consider an extension of the logit model which allows for repeated measurements of the covariates.

In order to develop the logit model, suppose that the survival time is known to have occurred in one of K intervals. That is, we assume we begin a study with n subjects who are followed over time and we record the time interval where death occurred. If the subject was alive at last observation, we record the interval of this final observation. These latter observations are said to be "censored." As has been the case throughout this text, we let x denote the vector of covariates. This vector may contain treatment, risk factor, confounder, and interaction variables. Let Y_k be a variable denoting the occurrence or nonoccurrence of death in the kth interval. Let $\pi_k(\mathbf{x})$ denote the conditional probability that death occurs in the kth interval, conditional that death did not occur prior to the kth interval in a subject with covariate vector x. Specifically,

$$\pi_k(\mathbf{x}) = P(Y_k = 1 \mid Y_{k'} = 0 \text{ for } k' < k, \mathbf{x})$$

The logit model assumes that $g_k(\mathbf{x}) = \text{logit}[\pi_k(\mathbf{x})]$ is the linear function of the covariates, $g_k(\mathbf{x}) = \alpha_k + \text{ß}'\mathbf{x}$, specifically

$$\pi_k(\mathbf{x}) = \frac{e^{\alpha_k + \text{ß}'\mathbf{x}}}{1 + e^{\alpha_k + \text{ß}'\mathbf{x}}} \tag{8.11}$$

The parameter α_k corresponds to logit for a subject with $\mathbf{x} = \mathbf{0}$ in the kth interval. As in logistic regression, this intercept parameter will not be of interest unless we wish to estimate the conditional probabilities. In this section the vector ß contains only the p slope coefficients, $\text{ß}' = (\beta_1, \beta_2, ..., \beta_p)$.

We now consider development of the likelihood function for the logit model. Let $\mathbf{x}_i$, $i = 1, 2, ..., n$, denote the vector of covariates for the ith subject. We extend the definition of the outcome variable Y to include observations on each subject and for each interval the subject was observed. Denote the observed values of this variable as y_{ik} where $y_{ik} = 1$ if the ith subject died in the kth interval and let $y_{ik} = 0$ if the ith subject was known to be alive in the kth interval. It may be shown that the ith subject's contribution to the likelihood function is

$$l_i(\alpha, \text{ß}) = \pi_k(\mathbf{x}_i)^{y_{ik}} \prod_{k'=1}^{k-1} [1 - \pi_{k'}(\mathbf{x}_i)]^{1-y_{ik'}}$$

if the subject dies in the kth interval, $y_{ik} = 1$, and

$$l_i(\alpha, \beta) = \prod_{k'=1}^{k} [1 - \pi_k(x_i)]^{1-y_{ik'}}$$

if the subject is censored in the kth interval, and $y_{ik'} = 0$ for $k' = 1, 2, ..., k$.

In the likelihood function $\alpha' = (\alpha_1, \alpha_2, ..., \alpha_K)$ denotes the vector of interval-specific intercept parameters. The full likelihood is obtained as the product of $l_i(\alpha, \beta)$ over the n subjects and is

$$L(\alpha, \beta) = \prod_{i=1}^{n} l_i(\alpha, \beta)$$

This likelihood is in the form of a logistic regression model where each subject contributes k terms. With the modification in the data to be described we may use standard logistic regression software to obtain maximum likelihood estimates and estimates of the standard errors of the estimated parameters.

We illustrate the method of modifying the data set assuming that there are $K = 4$ intervals. The extension of the procedure to other numbers of intervals is not difficult. To include the interval specific parameters, α, we need to create, in the data, the value of an interval variable. For sake of ease of notation we will denote this by k. In Table 8.12 we illustrate the contribution of a subject to the new data set for each possible occurrence of the event.

For example, a subject who died in the third interval would contribute three records, or lines of data, to the file. In the first line the value of the interval variable would be 1 ($k = 1$), the value of the outcome variable would be zero ($y = 0$), and x denotes the value of the vector of covariates for this subject. In the second line of data the value of the interval variable would be 2 ($k = 2$), the value of the outcome variable would be 0 ($y = 0$), and the value of the vector of covariates would be the same as the first line, x. In the third line the value of the interval variable is 3 ($k = 3$) and the value of the outcome variable is now 1 ($y = 1$) since the subject died in this interval and the value of the vector of covariates would again be repeated as x. If the subject was censored in the third interval we assume the subject was alive for the entire interval and censoring took place at the right-hand endpoint. Hence, this subject would also contribute three lines of data to the file. In the first line the value of the interval variable would be 1 ($k = 1$), the value of the outcome variable would be 0 ($y = 0$) and the value of the vector of covariates would be x. In the second line of data the value of the interval variable would be 2 ($k = 2$), the value of the outcome variable would be 0 ($y = 0$) and the value of the vector of covariates would be the same as the first line, x. In the third line the value of the inerval variable would be 3 ($k = 3$), the outcome variable would be equal to zero ($y = 0$), and the covariates the same as lines 1 and 2. This pattern of contribution of lines of data is similar for observations occurring in the the other intervals. Note that subjects who have been censored in the first interval do not contribute any data to the analysis.

Table 8.12 Contribution to a Modified Data File Listed by the Interval Where Death or Censoring Occurred and the Status of the Subject in That Interval.

Interval Where Death or Censoring Occurred	Status in Interval	Value of Interval Variable (k)	Value of the Outcome Variable (y)	Value of the Covariates
1	Dead	1	1	x
	Censored	1	0	x
2	Dead	1	0	x
		2	1	x
	Censored	1	0	x
		2	0	x
3	Dead	1	0	x
		2	0	x
		3	1	x
	Censored	1	0	x
		2	0	x
		3	0	x
4	Dead	1	0	x
		2	0	x
		3	0	x
		4	1	x
	Censored	1	0	x
		2	0	x
		3	0	x
		4	0	x

To obtain the maximum likelihood estimates of the interval specific parameters, α_k, from standard logistic regression software we would declare the interval variable to be categorical with K levels and include among our covariates the constant $x_0 = 1$. If we use reference cell parameterization for the design variables for the interval variable, with the first interval as the reference cell, then the estimate of the β_0 would correspond to the parameter estimate for the first interval, $\hat{\alpha}_1 = \hat{\beta}_0$. The estimate of the parameter for the kth interval, $k > 1$, would be obtained as the sum of the estimate of β_0 and the estimate of the coefficient for the k–first design variable for interval, $\hat{\alpha}_k = \hat{\beta}_0 + \hat{\gamma}_{k-1}$. We use $\hat{\gamma}$ to denote the estimate of the coefficient of the design variable for interval to differentiate it from the interval-specific and slope parameters.

Note that the need to include interval-specific parameters points out a potential limitation of this model. Namely, if we use too many intervals then we may increase the number of parameters to a point where there are of the same order of magnitude as the sample size. This situation is similar to the one discussed in Chapters 6 and 7 in the context of stratified likelihoods for matched case-control studies. If this situation occurs then we should consider a regression model for continuous time such as the proportional hazards model.

The estimate of the conditional probability that a subject with covariates $\mathbf{x}$ will die in the kth interval given that the subject was alive through the k–first interval, for a subject with covariates $\mathbf{x}$, is

$$\hat{\pi}_k(\mathbf{x}) = \frac{e^{\hat{\alpha}_k + \hat{\beta}'\mathbf{x}}}{1 + e^{\hat{\alpha}_k + \hat{\beta}'\mathbf{x}}} \tag{8.12}$$

The estimate of the probability that a subject with covariates $\mathbf{x}$ survives from the beginning of the first interval through the kth interval is

$$\prod_{k'=1}^{k} \hat{\pi}_{k'}(\mathbf{x})$$

The analysis has been presented up to this point as if the covariates represent those in a final model. Methods for model building to reach this point are identical to those described in Chapter 4 with the exception that the design variables for the interval variable must appear in any model. Hence, the intercept only model contains γ and β_0. The interpretation of the estimated slope coefficients is the same as that described in Chapter 2 except that the relevant probability in the log-odds ratio is the conditional probability $\hat{\pi}_k(\mathbf{x})$. Interval estimates for parameters may be obtained using the estimated standard errors from the logistic regression software.

It is not clear how or if it is possible at all to use diagnostics computed by logistic regression software to assess the fit of the model. The problem arises from the fact we have adapted logistic regression software to fit a model which is, in reality, not the same logistic regression model the software was designed to handle. Also, each subject may contribute more than one line of data to the analysis and the software assumes each line corresponds to an independent subject. Summary statistics such as the deviance, as computed in logistic regression software, may not be meaningful. One way to deal with these problems (though not necessarily recommended) is to use a single interval, $K = 1$.

While using a single interval may simplify the computations there are several concerns. If we use a single interval then we have made the tacit assumption that $\alpha_1 = \alpha_k$. This assumption could be tested via a likelihood ratio test which excludes k, the interval variable from the full k interval analysis. Also there is undoubtedly a loss of efficiency in the estimates of the slope coefficients. This situation is somewhat analogous to estimating a mean from grouped data with very broad groups. Also if we use a single interval then, as noted above, all censored subjects (those with $y = 0$) must have been followed for the full length of the interval, however it is defined. Hauck (1985) gives a good example of what can happen if this condition is not followed. For example, if the interval of interest is 2 years then the analysis would be restricted to subjects whose survival time was less than 2 years and those subjects who were alive or censored after 2 years of study. Subjects censored within the first two years would have to be excluded from the analysis.

In summary, a method has been described for using conventional logistic regression software to estimate the coefficients in the discrete logistic failure time model. These estimates will be similar to those obtained from an analysis based on the proportional hazards model when the event rates are small in each interval, nominally less than 0.1, and the follow-up period is relatively short and broken into a few intervals. If a large number of intervals are required then an analysis based on a regression model for continuous time such as the proportional hazards model may be more appropriate.

Exercises

1. Appendix 6 contains the data from the mammography experience study described in Section 8.1. Use a subset of these data and fit a polytomous logistic regression model. For example, you may choose to use only the first 200 subjects. The purpose of the exercise is to obtain practice when there are more than two categories of outcome. Hence, any alternative strategy for identifying a subset of subjects is acceptable.

2. The data for the low birth weight study are listed in Appendix 1 and a code sheet is given in Table 4.1. We may use the data on birth weight (BWT) to define an outcome variable, y, with three levels as follows: If BWT $\geq$ 2500 then $y = 1$, if $2000 \leq$ BWT < 2500 then $y = 2$, and if BWT < 2000 then $y = 3$. Use the outcome variable y and fit a polytomous logistic regression model. In each of the above problems the steps in fitting the model should include: (1) a complete univariate analysis, (2) an appropriate selection of variables for a multivariate model (this should include scale identification for continuous covariates and assessment of the need for interactions), (3) an assessment of fit of the multivariate model, (4) preparation and presentation of a table containing the results of the final model (this table should contain point and interval estimates for all relevant odds ratios), and (5) conclusions from the analysis.

3. Using the final models identified in problems 1 and 2 compare the estimates of the coefficients obtained from fitting via polytomous logistic regression program to those obtained by fitting via individualized logistic regressions. Note that these two fits may have been done as part of the assessment of fit of the polytomous models.

Appendix 1   Data for a Study of Risk Factors Associated with Low Infant
Birth Weight.   Data were collected at Baystate Medical Center, Springfield,
Massachusetts, during 1986.

ID	LOW	AGE	LWT	RACE	SMOKE	PTL	HT	UI	FTV	BWT
85	0	19	182	2	0	0	0	1	0	2523
86	0	33	155	3	0	0	0	0	3	2551
87	0	20	105	1	1	0	0	0	1	2557
88	0	21	108	1	1	0	0	1	2	2594
89	0	18	107	1	1	0	0	1	0	2600
91	0	21	124	3	0	0	0	0	0	2622
92	0	22	118	1	0	0	0	0	1	2637
93	0	17	103	3	0	0	0	0	1	2637
94	0	29	123	1	1	0	0	0	1	2663
95	0	26	113	1	1	0	0	0	0	2665
96	0	19	95	3	0	0	0	0	0	2722
97	0	19	150	3	0	0	0	0	1	2733
98	0	22	95	3	0	0	1	0	0	2750
99	0	30	107	3	0	1	0	1	2	2750
100	0	18	100	1	1	0	0	0	0	2769
101	0	18	100	1	1	0	0	0	0	2769
102	0	15	98	2	0	0	0	0	0	2778
103	0	25	118	1	1	0	0	0	3	2782
104	0	20	120	3	0	0	0	1	0	2807
105	0	28	120	1	1	0	0	0	1	2821
106	0	32	121	3	0	0	0	0	2	2835
107	0	31	100	1	0	0	0	1	3	2835
108	0	36	202	1	0	0	0	0	1	2836
109	0	28	120	3	0	0	0	0	0	2863
111	0	25	120	3	0	0	0	1	2	2877
112	0	28	167	1	0	0	0	0	0	2877
113	0	17	122	1	1	0	0	0	0	2906
114	0	29	150	1	0	0	0	0	2	2920
115	0	26	168	2	1	0	0	0	0	2920
116	0	17	113	2	0	0	0	0	1	2920
117	0	17	113	2	0	0	0	0	1	2920
118	0	24	90	1	1	1	0	0	1	2948
119	0	35	121	2	1	1	0	0	1	2948
120	0	25	155	1	0	0	0	0	1	2977

Appendix 1   (Continued)

ID	LOW	AGE	LWT	RACE	SMOKE	PTL	HT	UI	FTV	BWT
121	0	25	125	2	0	0	0	0	0	2977
123	0	29	140	1	1	0	0	0	2	2977
124	0	19	138	1	1	0	0	0	2	2977
125	0	27	124	1	1	0	0	0	0	2992
126	0	31	215	1	1	0	0	0	2	3005
127	0	33	109	1	1	0	0	0	1	3033
128	0	21	185	2	1	0	0	0	2	3042
129	0	19	189	1	0	0	0	0	2	3062
130	0	23	130	2	0	0	0	0	1	3062
131	0	21	160	1	0	0	0	0	0	3062
132	0	18	90	1	1	0	0	1	0	3076
133	0	18	90	1	1	0	0	1	0	3076
134	0	32	132	1	0	0	0	0	4	3080
135	0	19	132	3	0	0	0	0	0	3090
136	0	24	115	1	0	0	0	0	2	3090
137	0	22	85	3	1	0	0	0	0	3090
138	0	22	120	1	0	0	1	0	1	3100
139	0	23	128	3	0	0	0	0	0	3104
140	0	22	130	1	1	0	0	0	0	3132
141	0	30	95	1	1	0	0	0	2	3147
142	0	19	115	3	0	0	0	0	0	3175
143	0	16	110	3	0	0	0	0	0	3175
144	0	21	110	3	1	0	0	1	0	3203
145	0	30	153	3	0	0	0	0	0	3203
146	0	20	103	3	0	0	0	0	0	3203
147	0	17	119	3	0	0	0	0	0	3225
148	0	17	119	3	0	0	0	0	0	3225
149	0	23	119	3	0	0	0	0	2	3232
150	0	24	110	3	0	0	0	0	0	3232
151	0	28	140	1	0	0	0	0	0	3234
154	0	26	133	3	1	2	0	0	0	3260
155	0	20	169	3	0	1	0	1	1	3274
156	0	24	115	3	0	0	0	0	2	3274
159	0	28	250	3	1	0	0	0	6	3303
160	0	20	141	1	0	2	0	1	1	3317
161	0	22	158	2	0	1	0	0	2	3317
162	0	22	112	1	1	2	0	0	0	3317

Appendix 1 (Continued)

ID	LOW	AGE	LWT	RACE	SMOKE	PTL	HT	UI	FTV	BWT
163	0	31	150	3	1	0	0	0	2	3321
164	0	23	115	3	1	0	0	0	1	3331
166	0	16	112	2	0	0	0	0	0	3374
167	0	16	135	1	1	0	0	0	0	3374
168	0	18	229	2	0	0	0	0	0	3402
169	0	25	140	1	0	0	0	0	1	3416
170	0	32	134	1	1	1	0	0	4	3430
172	0	20	121	2	1	0	0	0	0	3444
173	0	23	190	1	0	0	0	0	0	3459
174	0	22	131	1	0	0	0	0	1	3460
175	0	32	170	1	0	0	0	0	0	3473
176	0	30	110	3	0	0	0	0	0	3475
177	0	20	127	3	0	0	0	0	0	3487
179	0	23	123	3	0	0	0	0	0	3544
180	0	17	120	3	1	0	0	0	0	3572
181	0	19	105	3	0	0	0	0	0	3572
182	0	23	130	1	0	0	0	0	0	3586
183	0	36	175	1	0	0	0	0	0	3600
184	0	22	125	1	0	0	0	0	1	3614
185	0	24	133	1	0	0	0	0	0	3614
186	0	21	134	3	0	0	0	0	2	3629
187	0	19	235	1	1	0	1	0	0	3629
188	0	25	95	1	1	3	0	1	0	3637
189	0	16	135	1	1	0	0	0	0	3643
190	0	29	135	1	0	0	0	0	1	3651
191	0	29	154	1	0	0	0	0	1	3651
192	0	19	147	1	1	0	0	0	0	3651
193	0	19	147	1	1	0	0	0	0	3651
195	0	30	137	1	0	0	0	0	1	3699
196	0	24	110	1	0	0	0	0	1	3728
197	0	19	184	1	1	0	1	0	0	3756
199	0	24	110	3	0	1	0	0	0	3770
200	0	23	110	1	0	0	0	0	1	3770
201	0	20	120	3	0	0	0	0	0	3770
202	0	25	241	2	0	0	1	0	0	3790
203	0	30	112	1	0	0	0	0	1	3799
204	0	22	169	1	0	0	0	0	0	3827

LOW BIRTH WEIGHT DATA

Appendix 1 (Continued)

ID	LOW	AGE	LWT	RACE	SMOKE	PTL	HT	UI	FTV	BWT
205	0	18	120	1	1	0	0	0	2	3856
206	0	16	170	2	0	0	0	0	4	3860
207	0	32	186	1	0	0	0	0	2	3860
208	0	18	120	3	0	0	0	0	1	3884
209	0	29	130	1	1	0	0	0	2	3884
210	0	33	117	1	0	0	0	1	1	3912
211	0	20	170	1	1	0	0	0	0	3940
212	0	28	134	3	0	0	0	0	1	3941
213	0	14	135	1	0	0	0	0	0	3941
214	0	28	130	3	0	0	0	0	0	3969
215	0	25	120	1	0	0	0	0	2	3983
216	0	16	95	3	0	0	0	0	1	3997
217	0	20	158	1	0	0	0	0	1	3997
218	0	26	160	3	0	0	0	0	0	4054
219	0	21	115	1	0	0	0	0	1	4054
220	0	22	129	1	0	0	0	0	0	4111
221	0	25	130	1	0	0	0	0	2	4153
222	0	31	120	1	0	0	0	0	2	4167
223	0	35	170	1	0	1	0	0	1	4174
224	0	19	120	1	1	0	0	0	0	4238
225	0	24	116	1	0	0	0	0	1	4593
226	0	45	123	1	0	0	0	0	1	4990
4	1	28	120	3	1	1	0	1	0	709
10	1	29	130	1	0	0	0	1	2	1021
11	1	34	187	2	1	0	1	0	0	1135
13	1	25	105	3	0	1	1	0	0	1330
15	1	25	85	3	0	0	0	1	0	1474
16	1	27	150	3	0	0	0	0	0	1588
17	1	23	97	3	0	0	0	1	1	1588
18	1	24	128	2	0	1	0	0	1	1701
19	1	24	132	3	0	0	1	0	0	1729
20	1	21	165	1	1	0	1	0	1	1790
22	1	32	105	1	1	0	0	0	0	1818
23	1	19	91	1	1	2	0	1	0	1885
24	1	25	115	3	0	0	0	0	0	1893
25	1	16	130	3	0	0	0	0	1	1899
26	1	25	92	1	1	0	0	0	0	1928

Appendix 1 (Continued)

ID	LOW	AGE	LWT	RACE	SMOKE	PTL	HT	UI	FTV	BWT
27	1	20	150	1	1	0	0	0	2	1928
28	1	21	200	2	0	0	0	1	2	1928
29	1	24	155	1	1	1	0	0	0	1936
30	1	21	103	3	0	0	0	0	0	1970
31	1	20	125	3	0	0	0	1	0	2055
32	1	25	89	3	0	2	0	0	1	2055
33	1	19	102	1	0	0	0	0	2	2082
34	1	19	112	1	1	0	0	1	0	2084
35	1	26	117	1	1	1	0	0	0	2084
36	1	24	138	1	0	0	0	0	0	2100
37	1	17	130	3	1	1	0	1	0	2125
40	1	20	120	2	1	0	0	0	3	2126
42	1	22	130	1	1	1	0	1	1	2187
43	1	27	130	2	0	0	0	1	0	2187
44	1	20	80	3	1	0	0	1	0	2211
45	1	17	110	1	1	0	0	0	0	2225
46	1	25	105	3	0	1	0	0	1	2240
47	1	20	109	3	0	0	0	0	0	2240
49	1	18	148	3	0	0	0	0	0	2282
50	1	18	110	2	1	1	0	0	0	2296
51	1	20	121	1	1	1	0	1	0	2296
52	1	21	100	3	0	1	0	0	4	2301
54	1	26	96	3	0	0	0	0	0	2325
56	1	31	102	1	1	1	0	0	1	2353
57	1	15	110	1	0	0	0	0	0	2353
59	1	23	187	2	1	0	0	0	1	2367
60	1	20	122	2	1	0	0	0	0	2381
61	1	24	105	2	1	0	0	0	0	2381
62	1	15	115	3	0	0	0	1	0	2381
63	1	23	120	3	0	0	0	0	0	2395
65	1	30	142	1	1	1	0	0	0	2410
67	1	22	130	1	1	0	0	0	1	2410
68	1	17	120	1	1	0	0	0	3	2414
69	1	23	110	1	1	1	0	0	0	2424
71	1	17	120	2	0	0	0	0	2	2438
75	1	26	154	3	0	1	1	0 ·	1	2442
76	1	20	105	3	0	0	0	0	3	2450

Appendix 1 (Continued)

ID	LOW	AGE	LWT	RACE	SMOKE	PTL	HT	UI	FTV	BWT
77	1	26	190	1	1	0	0	0	0	2466
78	1	14	101	3	1	1	0	0	0	2466
79	1	28	95	1	1	0	0	0	2	2466
81	1	14	100	3	0	0	0	0	2	2495
82	1	23	94	3	1	0	0	0	0	2495
83	1	17	142	2	0	0	1	0	0	2495
84	1	21	130	1	1	0	1	0	3	2495

Appendix 2 Data for a Study of Risk Factors Associated with ICU Mortality. Data were collected at Baystate Medical Center, Springfield, Massachusetts.

ID	STA	AGE	SEX	RACE	SER	CAN	CRN	INF	CPR	SYS	HRA	PRE	TYP	FRA	PO2	PH	PCO	BIC	CRE	LOC
8	0	27	1	1	0	0	0	1	0	142	88	0	1	0	0	0	0	0	0	0
12	0	59	0	1	0	0	0	0	0	112	80	1	1	0	0	0	0	0	0	0
14	0	77	0	1	1	0	0	0	0	100	70	0	0	0	0	0	0	0	0	0
28	0	54	0	1	0	0	0	1	0	142	103	0	1	1	0	0	0	0	0	0
32	0	87	1	1	1	0	0	1	0	110	154	1	1	0	0	0	0	0	0	0
38	0	69	0	1	0	0	0	1	0	110	132	0	1	0	1	0	0	1	0	0
40	0	63	0	1	1	0	0	0	0	104	66	0	0	0	0	0	0	0	0	0
41	0	30	1	1	1	0	0	0	0	144	110	0	1	0	0	0	0	0	0	0
42	0	35	0	2	0	0	0	0	0	108	60	0	1	0	0	0	0	0	0	0
50	0	70	1	1	1	0	0	0	0	138	103	1	0	0	0	0	0	0	0	0
51	0	55	1	1	1	1	0	1	0	188	86	0	0	0	0	0	0	0	0	0
53	0	48	0	2	1	0	0	0	0	162	100	0	0	0	0	0	0	0	0	0
58	0	66	1	1	1	0	1	0	0	160	80	0	1	0	0	0	0	0	0	0
61	0	61	1	1	1	0	0	0	0	174	99	0	0	0	0	1	1	1	1	0
73	0	66	0	1	0	0	0	0	0	206	90	1	1	0	0	0	0	0	0	0
75	0	52	0	1	1	0	0	1	0	150	71	0	0	0	0	0	0	1	1	0
82	0	55	0	1	1	0	0	1	0	140	116	0	0	0	0	0	0	0	0	0
84	0	59	0	1	1	0	0	1	0	48	39	1	1	0	1	0	1	1	0	2
92	0	63	0	1	0	0	0	0	0	132	128	1	1	0	0	0	0	0	0	0
96	0	72	0	1	1	0	0	0	0	120	80	1	0	0	0	0	0	0	0	0
98	0	60	0	1	0	0	0	1	1	114	110	0	1	0	0	0	0	0	0	0
100	0	78	0	1	1	0	0	0	0	180	75	0	0	0	0	0	0	0	0	0

Appendix 2 (Continued)

ID	STA	AGE	SEX	RACE	SER	CAN	CRN	INF	CPR	SYS	HRA	PRE	TYP	FRA	PO2	PH	PCO	BIC	CRE	LOC
102	0	16	1	1	0	0	0	0	0	104	111	0	1	0	0	0	0	0	0	0
111	0	62	0	1	1	0	1	0	0	200	120	0	0	0	0	0	0	0	0	0
112	0	61	0	1	0	0	0	1	0	110	120	0	1	0	0	0	0	0	0	0
136	0	35	0	1	0	0	0	0	0	150	98	0	1	0	0	0	1	0	0	0
137	0	74	1	1	1	0	0	0	0	170	92	0	0	0	0	0	0	0	0	0
143	0	68	0	1	1	0	0	0	0	158	96	0	0	0	0	0	0	0	0	0
153	0	69	1	1	1	0	0	0	0	132	60	0	1	0	0	0	0	0	0	0
170	0	51	0	1	0	0	0	0	0	110	99	0	1	0	0	0	0	0	0	0
173	0	55	0	1	1	0	0	1	0	128	92	1	1	0	0	0	0	0	0	0
180	0	64	1	3	1	0	0	1	0	158	90	1	1	0	0	0	0	0	0	0
184	0	88	1	1	1	0	0	1	0	140	88	0	1	1	0	0	0	0	0	0
186	0	23	1	1	1	1	0	0	0	112	64	0	1	0	0	0	0	0	0	0
187	0	73	1	1	1	0	0	0	0	134	60	1	0	0	0	0	0	0	0	0
190	0	53	0	3	1	0	0	0	0	110	70	0	0	0	0	0	0	0	0	0
191	0	74	0	1	1	0	0	0	0	174	86	0	0	0	0	0	0	0	0	0
207	0	68	0	1	1	0	0	0	0	142	89	0	0	0	0	0	0	0	0	0
211	0	66	1	1	0	0	0	1	0	170	95	1	1	0	0	0	0	0	0	0
214	0	60	0	1	1	1	0	1	0	110	92	0	0	0	0	0	0	0	0	0
219	0	64	0	1	1	0	0	1	0	160	120	0	0	0	0	0	0	0	0	0
225	0	66	0	2	1	1	0	1	0	150	120	0	0	0	0	0	0	0	0	0
237	0	19	1	1	1	0	0	1	0	142	106	0	1	0	0	0	1	0	0	0
247	0	18	1	1	0	0	0	0	0	146	112	0	1	0	0	0	0	0	0	0
249	0	63	0	1	1	0	0	1	0	162	84	1	1	0	0	0	0	0	0	0

Appendix 2 (Continued)

ID	STA	AGE	SEX	RACE	SER	CAN	CRN	INF	CPR	SYS	HRA	PRE	TYP	FRA	PO2	PH	PCO	BIC	CRE	LOC
260	0	45	0	1	0	0	0	0	0	126	110	0	1	0	0	0	0	0	0	0
266	0	64	0	1	0	0	0	0	0	162	114	0	1	0	0	0	0	0	0	0
271	0	68	1	1	0	0	0	1	0	200	170	1	1	0	0	0	0	0	0	0
276	0	64	1	1	0	0	0	1	0	126	122	0	1	1	1	1	1	0	0	0
277	0	82	0	1	1	0	0	0	0	135	70	0	0	0	0	0	0	0	0	0
278	0	73	0	1	1	0	0	0	0	170	88	0	1	0	0	0	0	0	0	0
282	0	70	0	1	0	0	0	0	0	86	153	1	1	0	0	0	0	0	0	0
292	0	61	0	1	1	0	1	1	0	68	124	0	1	0	0	0	0	1	0	0
295	0	64	0	1	1	0	1	1	0	116	88	0	0	0	0	0	0	0	0	0
297	0	47	0	1	1	0	1	1	0	120	83	0	0	0	0	0	0	0	0	0
298	0	69	0	1	1	0	0	0	0	170	100	0	1	0	0	0	0	0	0	0
308	0	67	1	1	0	0	1	1	0	190	125	0	0	0	0	0	0	0	0	0
310	0	18	0	1	1	0	1	1	0	156	99	0	0	0	0	0	0	0	0	0
319	0	77	0	1	1	0	0	1	0	158	107	0	0	0	0	0	0	0	0	0
327	0	32	0	2	1	0	0	0	0	120	84	0	1	0	0	0	0	0	0	0
333	0	19	1	1	0	0	0	1	0	104	121	1	0	0	0	0	0	0	0	0
335	0	72	1	1	0	0	0	0	0	130	86	0	1	0	0	0	0	0	0	0
343	0	49	0	1	0	0	0	1	0	112	112	0	1	0	0	0	0	0	0	0
357	0	68	1	1	1	0	1	1	0	154	74	0	0	0	0	0	0	0	0	0
362	0	82	0	1	1	1	0	1	1	130	131	0	1	0	0	0	0	0	0	0
365	0	32	1	3	0	0	0	1	0	110	118	0	1	0	0	0	0	0	0	0
369	0	78	1	1	1	0	0	1	0	126	96	0	1	0	0	0	0	1	0	0
370	0	57	0	1	0	0	0	1	0	128	104	0	1	0	0	0	0	0	0	0

Appendix 2 (Continued)

ID	STA	AGE	SEX	RACE	SER	CAN	CRN	INF	CPR	SYS	HRA	PRE	TYP	FRA	PO2	PH	PCO	BIC	CRE	LOC
371	0	46	1	1	1	1	0	0	0	132	90	0	1	0	0	0	0	0	0	0
376	0	23	0	1	0	0	0	1	0	144	88	0	1	0	0	0	0	0	0	0
378	0	55	0	1	0	0	0	0	0	132	112	0	1	1	0	0	0	0	0	0
379	0	18	0	1	1	0	0	0	0	112	76	0	1	1	0	0	0	0	0	0
381	0	20	0	1	1	0	0	0	0	164	108	0	1	0	0	0	0	0	0	0
382	0	75	1	1	1	0	0	0	0	100	48	0	0	0	0	0	0	0	0	0
398	0	79	0	1	1	0	0	1	0	112	67	0	0	0	0	0	0	0	0	0
401	0	40	0	1	1	0	0	0	0	140	65	0	1	1	0	0	0	0	0	0
409	0	76	0	1	1	0	0	1	0	110	70	0	1	0	0	0	0	0	0	0
413	0	66	1	1	1	0	0	1	0	139	92	0	0	0	0	0	0	0	0	0
416	0	76	0	1	0	0	0	1	0	190	100	0	1	0	0	0	0	0	0	0
438	0	80	1	1	1	0	0	0	0	162	44	0	1	0	0	0	0	0	0	0
439	0	23	1	1	0	0	0	1	0	120	88	0	1	0	0	0	0	0	0	0
440	0	48	0	2	1	0	0	1	0	92	162	1	1	0	0	0	0	0	0	0
455	0	67	0	2	1	0	0	0	0	90	92	1	0	0	0	0	0	0	0	0
462	0	69	1	1	1	0	0	0	0	150	85	0	1	0	0	0	0	0	0	0
495	0_	65	0	3	1	0	0	0	0	208	124	0	0	0	0	0	0	0	0	0
498	0	72	0	1	1	0	0	0	0	126	88	0	0	0	0	0	1	0	0	0
502	0	55	0	1	0	0	0	0	0	190	136	0	1	0	1	1	1	0	0	0
505	0	40	0	1	0	0	0	0	0	130	65	0	1	0	0	0	0	0	0	0
508	0	55	1	1	0	0	0	1	0	110	86	0	1	1	1	0	0	0	0	0
517	0	34	0	1	1	0	0	0	0	110	80	0	1	0	0	0	0	0	0	0
522	0	47	1	1	1	0	0	0	0	132	68	0	1	0	0	0	0	0	0	0

Appendix 2 (Continued)

ID	STA	AGE	SEX	RACE	SER	CAN	CRN	INF	CPR	SYS	HRA	PRE	TYP	FRA	PO2	PH	PCO	BIC	CRE	LOC
525	0	41	1	1	0	0	0	1	0	118	145	0	1	0	0	1	0	1	0	0
526	0	84	1	1	0	1	1	1	0	100	103	0	1	0	0	0	0	1	1	0
546	0	88	1	1	1	0	0	0	0	110	46	1	0	0	0	0	0	1	0	0
548	0	77	1	1	1	0	0	0	0	212	87	0	0	0	0	0	1	0	0	0
550	0	80	0	1	0	0	0	0	0	122	126	0	1	0	1	0	0	1	0	0
552	0	16	0	1	1	0	0	0	0	100	140	0	1	1	0	0	0	0	0	0
560	0	70	0	1	1	0	0	0	0	160	60	0	0	0	0	0	0	0	0	0
563	0	83	1	1	1	0	1	1	0	138	91	0	1	0	0	0	0	0	0	0
573	0	23	0	2	0	0	0	0	1	130	52	0	1	0	0	0	0	0	0	0
575	0	67	1	1	1	0	0	0	0	120	120	0	1	0	0	0	0	0	0	0
584	0	18	0	1	1	1	0	0	0	130	140	0	0	0	1	1	1	0	0	0
597	0	77	1	1	0	0	0	1	1	136	138	0	0	1	0	1	1	0	0	0
598	0	48	1	1	0	0	0	0	0	128	96	0	1	0	0	0	0	0	0	0
601	0	24	1	2	0	0	0	0	0	140	86	0	1	0	0	0	0	0	0	0
605	0	71	1	1	0	0	1	1	0	124	106	0	1	0	0	0	0	0	0	0
607	0	72	0	1	1	0	0	0	0	134	60	0	1	0	0	0	0	0	0	0
619	0	77	1	1	1	1	0	0	0	170	115	1	0	0	0	0	0	0	0	0
620	0	60	0	1	1	0	1	1	0	124	135	0	1	0	0	0	0	0	0	0
639	0	46	0	1	1	1	0	0	0	110	128	0	0	0	0	0	0	0	0	0
644	0	65	1	1	0	0	0	0	0	100	105	0	1	0	0	0	0	0	0	0
645	0	36	0	1	0	0	0	0	0	224	125	0	0	0	0	0	0	0	0	0
648	0	68	0	1	1	0	0	0	0	112	64	0	1	0	0	0	0	0	0	0
655	0	58	0	1	0	0	0	0	0	154	98	0	1	0	0	0	0	0	0	0

Appendix 2 (Continued)

ID	STA	AGE	SEX	RACE	SER	CAN	CRN	INF	CPR	SYS	HRA	PRE	TYP	FRA	PO2	PH	PCO	BIC	CRE	LOC
659	0	76	1	1	0	0	0	1	0	92	112	0	1	0	0	0	0	0	0	0
669	0	41	1	2	0	0	0	0	0	110	144	0	1	0	0	0	0	1	1	0
670	0	20	0	3	0	0	0	0	0	120	68	0	1	0	0	0	0	0	0	0
674	0	91	0	1	0	1	1	1	0	152	125	0	1	0	0	0	0	0	0	0
675	0	75	0	1	1	0	0	0	0	140	90	0	1	0	0	0	1	0	0	0
676	0	25	1	1	0	0	0	0	0	131	135	0	1	0	1	0	0	1	0	0
709	0	70	0	1	0	0	1	1	0	78	143	0	1	1	0	0	0	0	0	0
713	0	47	0	1	1	0	0	0	0	156	112	0	1	0	0	0	0	0	0	0
727	0	75	0	3	1	0	0	0	0	144	120	1	1	0	0	0	0	0	1	0
728	0	40	0	2	0	0	1	1	0	160	150	0	1	0	1	0	1	0	0	0
732	0	71	0	1	0	0	1	1	0	148	192	0	1	1	1	1	0	0	0	0
746	0	70	1	1	0	0	1	1	0	90	140	0	1	0	1	0	0	1	0	0
749	0	58	0	1	1	0	0	0	0	148	95	1	1	0	0	0	0	0	0	0
754	0	54	0	1	1	0	0	0	0	136	80	0	0	0	0	0	0	0	0	0
761	0	77	0	1	0	0	1	1	0	128	59	0	0	0	0	0	0	0	0	0
763	0	55	0	1	1	1	0	0	0	138	140	0	1	0	0	0	0	0	0	0
764	0	21	0	1	0	0	1	0	0	120	62	0	1	0	0	0	1	0	0	0
765	0	53	0	2	0	0	1	1	1	170	115	0	1	0	0	0	0	0	0	0
766	0	31	1	1	1	1	1	1	1	146	100	0	1	0	0	1	0	0	0	0
772	0	71	0	1	1	0	0	0	0	204	52	0	0	0	1	0	0	0	0	0
776	0	49	0	2	0	0	0	0	0	150	100	0	1	0	0	0	0	0	0	0
784	0	60	1	2	0	0	1	1	0	116	92	1	1	1	1	1	1	0	0	0
794	0	50	0	1	0	0	1	1	0	156	99	0	1	0	0	0	1	0	0	0

Appendix 2 (Continued)

ID	STA	AGE	SEX	RACE	SER	CAN	CRN	INF	CPR	SYS	HRA	PRE	TYP	FRA	PO2	PH	PCO	BIC	CRE	LOC
796	0	45	1	1	1	0	0	0	0	132	109	0	1	1	0	0	0	0	0	0
809	0	21	0	1	1	0	0	0	0	110	90	0	1	0	0	0	0	0	0	0
814	0	73	1	1	1	0	0	0	0	130	83	0	1	0	0	0	0	0	0	0
816	0	28	0	1	1	0	1	0	0	122	80	1	0	1	0	0	0	0	0	0
829	0	17	0	1	1	0	0	0	0	140	78	0	1	1	0	0	0	0	0	0
837	0	17	1	3	0	0	0	0	0	130	140	0	1	0	0	0	0	0	0	0
846	0	21	1	1	1	0	0	0	0	142	79	0	1	0	0	0	0	0	0	0
847	0	68	1	1	1	1	0	0	0	91	79	0	0	0	0	0	0	0	0	0
863	0	17	0	3	1	0	0	0	0	136	78	0	1	0	0	0	0	0	0	0
867	0	60	0	1	0	0	1	0	0	108	120	0	1	0	0	0	0	0	0	0
875	0	69	0	1	1	0	0	0	0	169	73	0	1	0	0	0	0	0	0	0
877	0	88	1	1	0	1	0	0	0	190	88	0	1	0	0	0	0	0	0	0
880	0	20	0	1	1	0	0	0	0	120	80	0	1	0	0	0	0	0	0	2
881	0	89	1	1	1	0	0	0	0	190	114	0	1	0	0	0	0	0	0	0
889	0	62	1	1	1	0	0	0	0	110	78	0	1	0	0	0	1	0	0	0
893	0	46	0	1	0	1	1	1	0	142	89	0	1	0	0	0	0	0	0	0
906	0	19	0	1	1	0	1	0	0	100	137	0	1	0	1	1	1	1	0	0
912	0	71	0	1	0	0	1	0	0	124	124	0	1	0	0	0	0	0	0	0
915	0	67	0	1	1	0	0	0	0	152	78	0	0	0	1	1	1	1	0	0
923	0	20	0	1	1	0	0	0	0	104	83	0	1	0	0	0	0	0	0	0
924	0	73	1	2	0	1	0	0	0	162	100	0	1	0	0	0	0	0	0	0
925	0	59	0	1	0	0	0	0	0	100	88	0	1	1	0	0	0	0	0	0
929	0	42	0	1	1	0	0	0	0	122	84	0	1	1	0	0	0	0	0	0

Appendix 2 (Continued)

ID	STA	AGE	SEX	RACE	SER	CAN	CRN	INF	CPR	SYS	HRA	PRE	TYP	FRA	PO2	PH	PCO	BIC	CRE	LOC
4	1	87	1	1	1	0	0	1	0	80	96	0	1	1	1	1	1	0	0	0
27	1	76	1	1	1	0	0	1	0	128	90	1	1	0	0	0	0	0	0	0
47	1	78	0	1	0	0	0	1	0	130	132	0	1	0	0	0	0	1	0	0
52	1	63	0	1	0	1	1	1	0	112	106	1	1	0	1	0	0	0	0	0
127	1	19	0	1	1	0	0	0	0	140	76	0	1	0	0	0	0	0	0	0
145	1	67	1	1	0	0	0	1	0	62	145	0	1	0	0	1	0	1	1	0
154	1	53	1	1	0	0	0	1	0	148	128	0	1	0	0	1	1	0	0	0
165	1	92	0	1	0	0	0	1	0	124	80	0	1	0	0	0	0	1	0	2
195	1	57	1	1	1	0	0	1	0	110	124	0	1	0	0	0	1	0	0	0
202	1	75	1	1	0	1	0	0	0	130	136	0	1	0	0	0	0	0	0	0
204	1	91	0	1	0	0	0	1	0	64	125	0	1	0	0	0	0	0	0	1
208	1	70	0	1	1	0	0	0	1	168	122	0	0	0	1	0	0	0	0	0
222	1	88	0	1	0	0	0	1	0	141	140	0	1	0	0	0	0	0	0	2
238	1	41	0	1	1	0	0	1	0	140	58	0	1	0	0	0	0	0	0	0
241	1	61	0	1	0	0	0	0	0	140	81	0	1	0	0	0	0	0	0	0
273	1	80	0	1	1	0	0	0	0	100	85	0	1	0	0	0	0	0	0	0
285	1	40	0	1	0	0	0	1	0	86	80	1	1	0	0	0	0	0	0	1
299	1	75	1	1	0	0	0	1	1	90	100	0	1	0	0	0	0	0	0	2
331	1	63	1	1	1	0	1	1	1	36	86	0	1	1	0	0	0	0	0	0
346	1	75	1	1	0	1	0	0	0	190	94	0	1	0	0	0	0	0	1	0
380	1	20	0	1	1	0	0	0	0	148	72	0	1	1	0	0	0	0	0	0
384	1	71	0	1	0	0	0	0	0	142	95	0	1	0	0	0	0	0	0	0
412	1	51	1	1	1	0	0	1	0	134	100	1	1	0	0	0	0	0	0	1

Appendix 2 (Continued)

ID	STA	AGE	SEX	RACE	SER	CAN	CRN	INF	CPR	SYS	HRA	PRE	TYP	FRA	PO2	PH	PCO	BIC	CRE	LOC
427	1	65	0	1	0	0	0	0	0	66	94	0	1	0	0	0	0	0	0	2
442	1	69	1	3	0	0	1	0	0	170	60	1	1	0	1	0	0	0	0	0
461	1	55	0	1	1	0	1	1	0	122	100	1	1	0	0	0	0	0	0	0
468	1	50	1	1	1	1	0	0	0	120	96	0	1	0	0	0	0	0	0	0
490	1	78	0	1	0	0	0	1	0	110	81	0	1	0	0	0	0	0	0	0
518	1	71	1	1	0	0	0	0	1	70	112	0	1	0	0	0	0	0	0	2
611	1	85	1	1	1	0	0	0	0	136	96	0	1	0	0	1	0	1	1	0
613	1	75	0	1	0	0	1	1	0	130	119	0	1	0	0	0	1	0	0	0
666	1	65	1	1	0	0	0	1	1	104	150	0	1	0	0	0	0	1	0	2
671	1	49	0	1	1	0	1	1	1	140	108	0	1	0	0	0	0	0	1	0
706	1	75	1	1	1	0	0	1	1	150	66	0	1	0	0	0	0	0	0	2
740	1	72	1	1	1	0	1	0	0	90	160	0	1	0	0	0	0	0	1	0
751	1	69	0	1	0	0	0	1	0	80	81	0	1	0	0	0	0	0	0	2
752	1	64	0	1	0	1	0	1	0	80	118	1	1	0	1	1	0	1	1	0
789	1	60	0	1	0	0	0	1	0	56	114	0	1	0	0	0	0	0	0	0
871	1	60	0	3	1	0	1	1	0	130	55	0	1	0	0	0	0	1	0	1
921	1	50	1	2	0	0	0	0	0	256	64	0	1	0	0	0	0	0	0	1

Appendix 3 Matched Case-Control Data Set. Pairs are selected from data in Appendix I. For each case a control is selected randomly after matching on age.

PAIR #	LOW	AGE	LWT	RACE	SMOKE	PTD	UT	UI
1	0	14	135	1	0	0	0	0
1	1	14	101	3	1	1	0	0
2	0	15	98	2	0	0	0	0
2	1	15	115	3	0	0	0	1
3	0	16	95	3	0	0	0	0
3	1	16	130	3	0	0	0	0
4	0	17	103	3	0	0	0	0
4	1	17	130	3	1	1	0	1
5	0	17	122	1	1	0	0	0
5	1	17	110	1	1	0	0	0
6	0	17	113	2	0	0	0	0
6	1	17	120	1	1	0	0	0
7	0	17	113	2	0	0	0	0
7	1	17	120	2	0	0	0	0
8	0	17	119	3	0	0	0	0
8	1	17	142	2	0	0	1	0
9	0	18	100	1	1	0	0	0
9	1	18	148	3	0	0	0	0
10	0	18	90	1	1	0	0	1
10	1	18	110	2	1	1	0	0
11	0	19	150	3	0	0	0	0
11	1	19	91	1	1	1	0	1
12	0	19	115	3	0	0	0	0
12	1	19	102	1	0	0	0	0
13	0	19	235	1	1	0	1	0
13	1	19	112	1	1	0	0	1
14	0	20	120	3	0	0	0	1
14	1	20	150	1	1	0	0	0
15	0	20	103	3	0	0	0	0
15	1	20	125	3	0	0	0	1
16	0	20	169	3	0	1	0	1
16	1	20	120	2	1	0	0	0
17	0	20	141	1	0	1	0	1
17	1	20	80	3	1	0	0	1
18	0	20	121	2	1	0	0	0
18	1	20	109	3	0	0	0	0

Appendix 3 (Continued)

PAIR #	LOW	AGE	LWT	RACE	SMOKE	PTD	UT	UI
19	0	20	127	3	0	0	0	0
19	1	20	121	1	1	1	0	1
20	0	20	120	3	0	0	0	0
20	1	20	122	2	1	0	0	0
21	0	20	158	1	0	0	0	0
21	1	20	105	3	0	0	0	0
22	0	21	108	1	1	0	0	1
22	1	21	165	1	1	0	1	0
23	0	21	124	3	0	0	0	0
23	1	21	200	2	0	0	0	0
24	0	21	185	2	1	0	0	0
24	1	21	103	3	0	0	0	0
25	0	21	160	1	0	0	0	0
25	1	21	100	3	0	1	0	0
26	0	21	115	1	0	0	0	0
26	1	21	130	1	1	0	1	0
27	0	22	95	3	0	0	1	0
27	1	22	130	1	1	0	0	0
28	0	22	158	2	0	1	0	0
28	1	22	130	1	1	1	0	1
29	0	23	130	2	0	0	0	0
29	1	23	97	3	0	0	0	1
30	0	23	128	3	0	0	0	0
30	1	23	187	2	1	0	0	0
31	0	23	119	3	0	0	0	0
31	1	23	120	3	0	0	0	0
32	0	23	115	3	1	0	0	0
32	1	23	110	1	1	1	0	0
33	0	23	190	1	0	0	0	0
33	1	23	94	3	1	0	0	0
34	0	24	90	1	1	1	0	0
34	1	24	128	2	0	1	0	0
35	0	24	115	1	0	0	0	0
35	1	24	132	3	0	0	1	0
36	0	24	110	3	0	0	0	0
36	1	24	155	1	1	1	0	0
37	0	24	115	3	0	0	0	0

Appendix 3 (Continued)

PAIR #	LOW	AGE	LWT	RACE	SMOKE	PTD	UT	UI
37	1	24	138	1	0	0	0	0
38	0	24	110	3	0	1	0	0
38	1	24	105	2	1	0	0	0
39	0	25	118	1	1	0	0	0
39	1	25	105	3	0	1	1	0
40	0	25	120	3	0	0	0	1
40	1	25	85	3	0	0	0	1
41	0	25	155	1	0	0	0	0
41	1	25	115	3	0	0	0	0
42	0	25	125	2	0	0	0	0
42	1	25	92	1	1	0	0	0
43	0	25	140	1	0	0	0	0
43	1	25	89	3	0	1	0	0
44	0	25	241	2	0	0	1	0
44	1	25	105	3	0	1	0	0
45	0	26	113	1	1	0	0	0
45	1	26	117	1	1	1	0	0
46	0	26	168	2	1	0	0	0
46	1	26	96	3	0	0	0	0
47	0	26	133	3	1	1	0	0
47	1	26	154	3	0	1	1	0
48	0	26	160	3	0	0	0	0
48	1	26	190	1	1	0	0	0
49	0	27	124	1	1	0	0	0
49	1	27	130	2	0	0	0	1
50	0	28	120	3	0	0	0	0
50	1	28	120	3	1	1	0	1
51	0	28	130	3	0	0	0	0
51	1	28	95	1	1	0	0	0
52	0	29	135	1	0	0	0	0
52	1	29	130	1	0	0	0	1
53	0	30	95	1	1	0	0	0
53	1	30	142	1	1	1	0	0
54	0	31	215	1	1	0	0	0
54	1	31	102	1	1	1	0	0
55	0	32	121	3	0	0	0	0

Appendix 3 (Continued)

PAIR #	LOW	AGE	LWT	RACE	SMOKE	PTD	UT	UI
55	1	32	105	1	1	0	0	0
56	0	34	170	1	0	1	0	0
56	1	34	187	2	1	0	1	0

Appendix 4 Matched Case-Control Data Set. Three controls per case selected randomly after matching on age from Appendix 1.

STR	OBS	AGE	LOW	LWT	SMOKE	HT	UI	PTD
1	1	16	1	130	0	0	0	0
1	2	16	0	112	0	0	0	0
1	3	16	0	135	1	0	0	0
1	4	16	0	95	0	0	0	0
2	1	17	1	130	1	0	1	1
2	2	17	0	103	0	0	0	0
2	3	17	0	122	1	0	0	0
2	4	17	0	113	0	0	0	0
3	1	17	1	120	0	0	0	0
3	2	17	0	113	0	0	0	0
3	3	17	0	119	0	0	0	0
3	4	17	0	119	0	0	0	0
4	1	18	1	148	0	0	0	0
4	2	18	0	100	1	0	0	0
4	3	18	0	90	1	0	1	0
4	4	18	0	229	0	0	0	0
5	1	18	1	110	1	0	0	1
5	2	18	0	107	1	0	1	0
5	3	18	0	100	1	0	0	0
5	4	18	0	90	1	0	1	0
6	1	19	1	91	1	0	1	1
6	2	19	0	138	1	0	0	0
6	3	19	0	189	0	0	0	0
6	4	19	0	147	1	0	0	0
7	1	19	1	102	0	0	0	0
7	2	19	0	150	0	0	0	0
7	3	19	0	235	1	1	0	0
7	4	19	0	184	1	1	0	0
8	1	19	1	112	1	0	1	0
8	2	19	0	182	0	0	1	0
8	3	19	0	95	0	0	0	0
8	4	19	0	132	0	0	0	0
9	1	20	1	150	1	0	0	0
9	2	20	0	120	0	0	1	0
9	3	20	0	105	1	0	0	0
9	4	20	0	141	0	0	1	1

Appendix 4 (Continued)

STR	OBS	AGE	LOW	LWT	SMOKE	HT	UI	PTD
10	1	20	1	120	1	0	0	0
10	2	20	0	103	0	0	0	0
10	3	20	0	127	0	0	0	0
10	4	20	0	170	1	0	0	0
11	1	20	1	121	1	0	1	1
11	2	20	0	169	0	0	1	1
11	3	20	0	121	1	0	0	0
11	4	20	0	120	0	0	0	0
12	1	21	1	200	0	0	1	0
12	2	21	0	108	1	0	1	0
12	3	21	0	124	0	0	0	0
12	4	21	0	185	1	0	0	0
13	1	21	1	100	0	0	0	1
13	2	21	0	160	0	0	0	0
13	3	21	0	110	1	0	1	0
13	4	21	0	115	0	0	0	0
14	1	22	1	130	1	0	1	1
14	2	22	0	85	1	0	0	0
14	3	22	0	130	1	0	0	0
14	4	22	0	125	0	0	0	0
15	1	22	1	130	1	0	0	0
15	2	22	0	120	0	1	0	0
15	3	22	0	112	1	0	0	1
15	4	22	0	169	0	0	0	0
16	1	23	1	97	0	0	1	0
16	2	23	0	130	0	0	0	0
16	3	23	0	119	0	0	0	0
16	4	23	0	123	0	0	0	0
17	1	23	1	110	1	0	0	1
17	2	23	0	128	0	0	0	0
17	3	23	0	190	0	0	0	0
17	4	23	0	110	0	0	0	0
18	1	24	1	132	0	1	0	0
18	2	24	0	115	0	0	0	0
18	3	24	0	115	0	0	0	0
18	4	24	0	110	0	0	0	0
19	1	24	1	138	0	0	0	0

Appendix 4 (Continued)

STR	OBS	AGE	LOW	LWT	SMOKE	HT	UI	PTD
19	2	24	0	90	1	0	0	1
19	3	24	0	133	0	0	0	0
19	4	24	0	116	0	0	0	0
20	1	25	1	85	0	0	1	0
20	2	25	0	118	1	0	0	0
20	3	25	0	125	0	0	0	0
20	4	25	0	120	0	0	0	0
21	1	25	1	92	1	0	0	0
21	2	25	0	120	0	0	1	0
21	3	25	0	140	0	0	0	0
21	4	25	0	241	0	1	0	0
22	1	25	1	105	0	0	0	1
22	2	25	0	155	0	0	0	0
22	3	25	0	95	1	0	1	1
22	4	25	0	130	0	0	0	0
23	1	26	1	190	1	0	0	0
23	2	26	0	113	1	0	0	0
23	3	26	0	168	1	0	0	0
23	4	26	0	160	0	0	0	0
24	1	28	1	120	1	0	1	1
24	2	28	0	140	0	0	0	0
24	3	28	0	250	1	0	0	0
24	4	28	0	134	0	0	0	0
25	1	28	1	95	1	0	0	0
25	2	28	0	120	1	0	0	0
25	3	28	0	120	0	0	0	0
25	4	28	0	130	0	0	0	0
26	1	29	1	130	0	0	1	0
26	2	29	0	150	0	0	0	0
26	3	29	0	135	0	0	0	0
26	4	29	0	130	1	0	0	0
27	1	30	1	142	1	0	0	1
27	2	30	0	107	0	0	1	1
27	3	30	0	153	0	0	0	0
27	4	30	0	137	0	0	0	0
28	1	31	1	102	1	0	0	1
28	2	31	0	100	0	0	1	0

Appendix 4 (Continued)

STR	OBS	AGE	LOW	LWT	SMOKE	HT	UI	PTD
28	3	31	0	150	1	0	0	0
28	4	31	0	120	0	0	0	0
29	1	32	1	105	1	0	0	0
29	2	32	0	121	0	0	0	0
29	3	32	0	132	0	0	0	0
29	4	32	0	134	1	0	0	1

Appendix 5   Matched Case-Control Data for a Study of Risk Factors Associated with Benign Breast Disease. Data collected in New Haven, Connecticut. Cases and controls are age matched at the time of interview.

STR	OBS	AGMT	FNDX	HIGD	DEG	CHK	AGP1	AGMN	NLV	LIV	WT	AGLP	MST
1	1	39	1	9	0	1	23	13	0	5	118	39	1
1	2	39	0	10	0	2	16	11	1	3	175	39	3
1	3	39	0	11	0	2	20	12	1	3	135	39	2
1	4	39	0	12	1	1	21	11	0	3	125	40	1
2	1	38	1	14	2	1	.	14	.	.	118	39	1
2	2	38	0	12	1	1	20	15	0	2	183	38	1
2	3	38	0	9	0	1	19	11	0	5	218	38	1
2	4	38	0	13	1	1	23	13	0	2	192	37	1
3	1	38	1	9	0	1	22	15	2	2	125	38	1
3	2	38	0	10	0	1	20	14	0	2	123	38	1
3	3	38	0	15	1	1	19	13	3	2	140	37	1
3	4	38	0	12	1	1	18	13	0	2	160	38	1
4	1	38	1	15	1	1	24	14	2	3	150	38	5
4	2	38	0	15	2	1	26	13	1	1	130	38	2
4	3	38	0	12	1	2	23	14	0	4	140	38	1
4	4	38	0	12	1	1	25	16	0	2	130	38	1
5	1	38	1	12	1	1	21	17	0	2	150	38	2
5	2	38	0	12	1	2	20	12	1	2	148	38	1
5	3	38	0	14	2	1	.	13	.	.	134	39	1
5	4	38	0	13	1	1	16	14	0	6	138	38	4
6	1	38	1	13	1	1	24	12	1	3	116	39	1
6	2	38	0	12	1	1	19	12	0	2	145	35	2

Appendix 5 (Continued)

STR	OBS	AGMT	FNDX	HIGD	DEG	CHK	AGP1	AGMN	NLV	LIV	WT	AGLP	MST
6	3	38	0	14	2	1	21	10	4	3	195	35	1
6	4	38	0	14	4	1	25	8	0	1	180	38	2
7	1	37	1	17	4	1	•	13	•	•	137	37	5
7	2	37	0	15	2	1	20	11	2	2	135	37	2
7	3	37	0	9	0	1	18	10	2	3	155	37	1
7	4	37	0	12	1	1	22	13	2	2	120	38	1
8	1	36	1	12	1	1	•	14	•	•	126	36	2
8	2	36	0	10	0	1	20	12	1	2	191	36	1
8	3	36	0	10	0	2	17	10	1	3	185	37	1
8	4	36	0	12	1	2	23	12	0	2	119	37	1
9	1	35	1	12	1	1	23	14	0	3	129	36	1
9	2	35	0	14	1	2	21	11	0	3	170	34	2
9	3	36	0	12	1	1	22	14	0	4	110	36	1
9	4	35	0	14	2	2	24	11	0	2	155	35	1
10	1	35	1	12	1	2	21	12	1	2	105	29	1
10	2	36	0	17	3	1	26	13	2	2	115	36	1
10	3	36	0	12	1	2	22	12	0	3	120	36	1
10	4	36	0	12	1	1	33	16	•	1	150	36	1
11	1	35	1	20	5	1	•	11	2	•	135	35	1
11	2	35	0	10	0	2	18	13	0	2	110	35	2
11	3	35	0	12	1	1	19	11	0	3	170	36	1
11	4	35	0	14	1	1	21	12	1	2	145	36	1
12	1	34	1	12	1	2	25	10	1	1	170	34	1

Appendix 5 (Continued)

STR	OBS	AGMT	FNDX	HIGD	DEG	CHK	AGP1	AGMN	NLV	LIV	WT	AGLP	MST
12	2	35	0	18	4	1	27	13	0	4	140	35	1
12	3	34	0	12	1	1	20	11	0	3	240	34	1
12	4	34	0	12	1	2	25	16	1	1	100	35	1
13	1	33	1	20	5	1	•	14	•	•	92	33	5
13	2	33	0	15	1	1	21	11	0	1	160	33	1
13	3	32	0	12	1	1	24	12	0	2	155	32	1
13	4	33	0	14	2	1	25	12	1	2	132	33	1
14	1	33	1	18	4	1	28	14	0	5	110	33	1
14	2	33	0	12	1	1	21	12	0	2	145	29	5
14	3	33	0	13	1	1	20	13	1	2	155	29	3
14	4	33	0	18	4	1	21	13	0	1	110	33	1
15	1	32	1	12	1	1	30	13	0	1	129	32	1
15	2	32	0	12	1	1	25	11	0	2	131	32	1
15	3	32	0	15	1	2	20	9	1	2	218	26	3
15	4	32	0	12	1	1	23	16	0	2	115	32	1
16	1	31	1	17	3	1	30	14	1	0	110	30	1
16	2	30	0	10	0	1	21	14	0	3	130	30	1
16	3	31	0	13	1	1	23	11	0	2	97	31	1
16	4	31	1	13	1	1	24	13	0	3	120	31	1
17	1	68	0	14	2	1	22	12	0	3	130	50	2
17	2	68	0	8	0	1	34	14	0	3	150	53	4
17	3	68	0	16	3	2	•	13	•	•	123	35	5
17	4	68	0	12	1	2	19	12	0	7	145	46	4

Appendix 5 (Continued)

STR	OBS	AGMT	FNDX	HIGD	DEG	CHK	AGP1	AGMN	NLV	LIV	WT	AGLP	MST
18	1	64	1	12	1	2	30	14	1	3	135	53	1
18	2	64	0	20	4	1	·	14	·	·	132	44	5
18	3	64	0	13	1	1	26	11	0	5	205	42	4
18	4	64	0	12	1	1	25	10	0	2	127	50	4
19	1	63	1	10	0	1	21	15	0	5	120	52	1
19	2	63	0	12	1	2	·	12	·	·	145	46	5
19	3	63	0	5	0	2	·	14	·	·	175	51	3
19	4	63	0	12	1	2	24	11	0	3	144	50	1
20	1	62	1	12	1	2	·	16	·	·	163	33	5
20	2	62	0	12	1	1	26	15	0	2	170	39	1
20	3	62	0	16	3	2	32	12	0	2	134	53	4
20	4	62	0	10	0	1	22	12	1	3	155	39	4
21	1	61	1	8	0	1	28	14	0	3	145	53	1
21	2	61	0	13	1	1	26	13	0	1	140	50	1
21	3	61	0	8	0	1	28	15	1	3	120	41	1
21	4	61	0	16	3	1	27	14	0	2	134	45	1
22	1	61	1	11	0	1	22	16	0	4	150	56	1
22	2	62	0	9	0	2	30	11	0	1	117	36	2
22	3	62	0	15	2	1	25	15	1	4	147	52	1
22	4	61	0	14	1	1	26	13	1	3	124	52	1
23	1	61	1	12	1	1	26	17	0	2	129	34	1
23	2	62	0	18	4	1	33	11	0	1	170	54	2
23	3	61	0	6	0	2	25	13	0	3	153	50	1

Appendix 5 (Continued)

STR	OBS	AGMT	FNDX	HIGD	DEG	CHK	AGP1	AGMN	NLV	LIV	WT	AGLP	MST
23	4	61	0	12	1	1	29	13	1	2	130	55	4
24	1	61	1	10	0	2	21	15	0	3	145	53	1
24	2	61	0	8	0	1	18	13	0	5	140	56	4
24	3	61	0	12	1	1	22	17	0	2	155	55	1
24	4	61	0	8	0	1	23	15	1	3	116	43	1
25	1	60	1	13	1	1	28	17	0	2	115	51	1
25	2	60	0	12	1	2	25	11	0	2	175	42	1
25	3	60	0	11	0	2	24	13	0	2	179	50	1
25	4	60	0	13	2	1	33	15	0	3	119	47	1
26	1	58	1	12	1	1	20	12	1	5	153	53	1
26	2	58	0	12	1	2	25	16	0	3	185	55	1
26	3	58	0	13	2	2	.	12	.	.	280	42	1
26	4	58	0	12	1	1	24	10	1	0	140	25	2
27	1	55	1	14	2	1	30	16	1	2	126	44	1
27	2	55	0	12	1	2	30	13	0	2	193	50	4
27	3	55	0	12	1	2	.	12	.	.	140	55	5
27	4	55	0	11	0	1	24	14	0	6	116	47	1
28	1	55	1	12	1	1	24	14	0	4	140	52	1
28	2	55	0	12	1	2	.	14	.	.	138	50	1
28	3	55	0	12	1	1	16	12	2	3	175	47	1
28	4	55	0	10	0	1	26	15	2	4	155	50	3
29	1	52	1	12	1	2	.	12	.	.	125	36	5
29	2	52	0	14	2	1	28	12	0	2	113	45	1

Appendix 5 (Continued)

STR	OBS	AGMT	FNDX	HIGD	DEG	CHK	AGP1	AGMN	NLV	LIV	WT	AGLP	MST
29	3	52	0	8	0	2	20	14	2	6	110	40	4
29	4	52	0	12	1	2	25	13	0	3	190	48	1
30	1	52	1	12	1	1	23	14	0	3	114	50	1
30	2	52	0	14	2	1	21	12	0	3	126	43	1
30	3	52	0	9	0	1	23	11	1	2	159	42	1
30	4	52	0	12	1	1	20	11	0	5	170	42	1
31	1	51	1	7	0	2	24	16	0	5	156	52	1
31	2	51	0	16	3	2	24	12	3	4	161	50	1
31	3	51	0	15	2	1	22	13	0	2	150	45	1
31	4	51	0	15	2	1	24	13	0	5	115	51	1
32	1	49	1	20	4	1	•	14	0	•	95	49	5
32	2	49	0	12	1	2	25	12	0	2	235	44	1
32	3	49	0	12	1	1	24	13	0	3	145	44	1
32	4	49	0	14	2	1	25	13	0	3	123	49	1
33	1	48	1	17	4	1	22	11	0	3	145	48	1
33	2	48	0	12	1	2	22	11	0	1	155	48	1
33	3	48	0	12	1	1	•	12	•	•	115	48	1
33	4	48	0	12	1	2	19	11	7	0	190	29	1
34	1	47	1	17	4	1	26	14	0	4	120	47	1
34	2	47	0	15	1	2	20	12	0	5	110	47	1
34	3	47	0	12	1	1	24	14	0	2	148	45	1
34	4	47	0	10	0	1	22	13	0	3	120	45	1
35	1	47	1	12	1	1	19	12	0	1	132	47	2

Appendix 5 (Continued)

STR	OBS	AGMT	FNDX	HIGD	DEG	CHK	AGP1	AGMN	NLV	LIV	WT	AGLP	MST
35	2	47	0	10	0	2	23	15	1	3	115	29	1
35	3	47	0	11	0	1	23	13	0	2	125	47	1
35	4	47	0	17	3	1	21	12	1	5	120	39	2
36	1	46	1	10	0	2	27	15	1	11	155	46	4
36	2	46	0	12	1	1	19	11	0	3	170	45	1
36	3	46	0	14	2	1	26	13	0	7	180	46	1
36	4	46	0	8	0	1	15	13	0	1	179	40	1
37	1	46	1	12	1	1	27	12	4	4	137	46	1
37	2	46	0	12	1	1	23	12	0	4	107	46	1
37	3	46	0	12	1	1	22	11	0	6	144	46	2
37	4	46	0	11	0	1	17	13	0	3	89	39	1
38	1	45	1	12	1	1	33	14	0	2	80	45	1
38	2	45	0	12	1	1	25	13	1	1	142	38	1
38	3	45	0	8	0	2	20	11	1	1	150	45	1
38	4	45	0	13	1	1	22	11	0	3	154	46	1
39	1	45	1	12	1	2	•	12	•	•	90	45	5
39	2	45	0	12	1	2	23	11	0	2	150	45	1
39	3	45	0	12	1	1	20	12	0	1	102	28	1
39	4	45	0	19	4	1	30	12	0	3	110	45	1
40	1	45	1	12	1	1	18	15	4	4	101	45	1
40	2	45	0	12	1	1	22	17	1	2	109	40	1
40	3	45	0	13	1	1	30	13	0	2	210	40	1
40	4	45	0	12	1	1	22	10	0	5	198	33	1

Appendix 5 (Continued)

STR	OBS	AGMT	FNDX	HIGD	DEG	CHK	AGP1	AGMN	NLV	LIV	WT	AGLP	MST
41	1	45	1	17	4	1	25	16	1	4	124	45	1
41	2	45	0	12	1	2	23	12	3	3	133	45	1
41	3	45	0	16	3	1	23	13	0	3	120	46	1
41	4	45	0	14	1	1	23	12	0	4	165	35	1
42	1	44	1	16	3	1	25	12	0	3	130	44	1
42	2	44	0	18	4	1	27	13	1	3	240	45	1
42	3	44	0	12	1	1	27	14	0	1	125	44	1
42	4	44	0	12	1	1	•	13	•	•	183	44	4
43	1	44	1	12	1	1	24	15	0	1	130	44	1
43	2	44	0	12	1	2	22	15	0	1	105	44	4
43	3	44	0	12	1	1	23	12	0	5	123	33	1
43	4	44	0	12	1	1	18	17	1	7	180	44	1
44	1	43	1	16	3	1	27	15	0	2	130	43	4
44	2	43	0	16	3	1	31	12	0	1	104	43	1
44	3	43	0	12	1	1	14	12	1	2	158	21	1
44	4	43	0	12	1	1	20	14	0	6	160	39	1
45	1	28	1	16	3	1	•	12	•	•	108	29	1
45	2	27	0	11	0	1	22	12	0	1	127	27	1
45	3	28	0	12	1	1	20	11	0	2	145	27	1
45	4	28	0	12	1	1	23	16	0	2	127	29	1
46	1	53	1	16	3	1	29	12	0	4	132	50	1
46	2	53	0	12	1	1	28	11	0	3	140	49	1
46	3	53	0	12	1	2	•	12	•	•	98	43	2

Appendix 5 (Continued)

STR	OBS	AGMT	FNDX	HIGD	DEG	CHK	AGP1	AGMN	NLV	LIV	WT	AGLP	MST
46	4	53	0	11	0	1	26	11	0	1	130	49	4
47	1	56	1	12	1	1	21	17	1	6	130	47	1
47	2	56	0	12	1	2	27	11	0	4	265	42	1
47	3	56	0	16	3	1	26	13	0	4	195	50	1
47	4	56	0	12	1	1	25	12	2	2	125	47	1
48	1	41	1	12	1	1	25	16	1	3	105	27	3
48	2	41	0	12	1	1	20	13	1	4	161	31	4
48	3	41	0	12	1	2	21	14	0	5	135	36	2
48	4	41	0	16	3	1	22	12	0	4	185	41	2
49	1	41	1	10	0	1	40	15	0	1	115	41	1
49	2	41	0	11	0	1	21	16	0	3	140	41	1
49	3	40	0	15	1	1	21	12	0	4	145	40	1
49	4	41	0	12	1	2	26	14	2	3	195	41	1
50	1	41	1	14	1	1	34	13	1	2	138	42	1
50	2	42	0	10	0	1	•	13	•	•	118	41	2
50	3	41	0	15	1	2	30	12	1	2	129	41	1
50	4	41	0	13	1	1	21	12	0	2	180	41	1

Appendix 6 Data Collected for a Study of
Factors Related to Mammography Experience.

OBS	ME	SYMPT	PB	HIST	BSE	DETC
1	3	3	7	0	1	2
2	3	2	11	0	1	1
3	3	3	8	1	1	1
4	1	3	11	0	1	1
5	2	4	7	0	1	1
6	3	3	7	0	1	1
7	2	4	6	0	1	2
8	3	4	6	0	1	1
9	3	2	6	0	1	1
10	1	4	6	0	1	1
11	3	4	8	0	1	2
12	3	3	6	1	0	1
13	3	4	6	0	1	1
14	3	1	5	1	1	1
15	3	2	8	0	0	2
16	3	4	11	0	1	1
17	3	3	6	0	1	1
18	1	4	5	0	1	1
19	3	3	10	0	1	1
20	3	3	10	0	1	1
21	2	3	5	0	1	1
22	3	4	5	0	1	2
23	3	1	5	0	1	1
24	3	2	8	0	1	2
25	1	2	9	1	1	2
26	2	4	7	0	1	1
27	3	4	7	0	1	2
28	3	3	10	0	1	1
29	2	3	7	0	1	2
30	3	3	7	0	1	1
31	1	4	7	0	1	2
32	2	2	5	0	1	1
33	2	3	8	0	1	1
34	3	1	5	0	1	1
35	3	2	9	0	1	1
36	2	3	11	0	1	1

Appendix 6 (Continued)

OBS	ME	SYMPT	PB	HIST	BSE	DETC
37	1	3	7	0	1	1
38	2	3	9	1	1	3
39	3	3	5	1	1	1
40	1	3	5	1	1	1
41	3	2	10	0	0	3
42	3	4	5	0	1	1
43	3	3	9	0	1	2
44	3	2	8	0	1	2
45	1	3	9	0	1	1
46	3	3	5	0	1	1
47	3	3	9	0	1	2
48	3	2	9	0	0	3
49	3	1	12	0	0	3
50	3	4	9	0	1	2
51	2	4	5	0	1	1
52	3	2	6	0	1	2
53	1	4	8	0	1	1
54	3	4	7	0	1	1
55	2	3	9	0	1	2
56	1	4	6	0	1	2
57	1	4	6	1	1	2
58	3	3	10	0	1	1
59	3	2	12	0	1	2
60	3	4	6	0	1	1
61	3	1	6	0	0	1
62	3	3	10	0	1	1
63	3	2	12	0	0	1
64	2	4	6	0	1	1
65	3	2	9	0	1	1
66	3	2	9	0	1	1
67	3	3	9	0	0	2
68	3	4	6	0	1	3
69	3	3	8	0	0	1
70	3	3	6	0	1	1
71	1	4	5	1	1	1
72	3	3	10	0	0	1
73	3	4	6	0	1	1

Appendix 6 (Continued)

OBS	ME	SYMPT	PB	HIST	BSE	DETC
74	3	1	6	0	1	2
75	1	4	5	0	1	1
76	1	4	5	0	1	1
77	3	2	6	0	1	2
78	1	3	5	0	1	2
79	3	2	7	0	1	1
80	1	1	6	0	0	2
81	2	4	7	0	1	1
82	1	1	6	0	1	1
83	1	4	7	1	0	1
84	1	4	6	0	1	1
85	3	4	5	0	1	1
86	1	3	9	0	1	1
87	2	2	6	0	1	1
88	3	2	11	0	0	1
89	3	4	9	0	0	1
90	2	1	9	0	0	3
91	3	2	7	0	0	2
92	3	3	10	0	1	1
93	1	4	8	0	1	1
94	1	3	6	0	1	1
95	3	3	7	0	1	1
96	1	3	8	1	1	1
97	3	2	10	0	1	1
98	3	3	7	1	1	1
99	3	1	9	0	1	1
100	2	4	6	0	1	1
101	2	4	8	0	1	1
102	3	2	8	0	1	1
103	3	4	8	0	1	1
104	2	4	6	0	1	1
105	3	2	10	0	1	2
106	2	4	5	0	1	1
107	3	4	5	0	1	1
108	3	3	5	1	1	1
109	1	3	7	0	1	1
110	3	3	10	0	1	1

Appendix 6 (Continued)

OBS	ME	SYMPT	PB	HIST	BSE	DETC
111	1	3	5	0	1	1
112	2	3	8	0	1	1
113	3	3	5	0	1	2
114	3	1	13	0	1	3
115	3	2	6	0	1	1
116	2	4	5	0	0	2
117	1	3	6	0	1	1
118	3	4	9	0	1	1
119	3	1	5	0	1	2
120	3	4	6	0	0	1
121	1	4	10	0	1	1
122	1	4	7	0	1	1
123	3	4	7	0	0	1
124	2	3	7	0	1	2
125	3	4	5	0	1	1
126	3	1	5	0	1	1
127	2	2	7	1	1	1
128	2	3	11	0	1	3
129	1	4	6	0	1	1
130	3	1	8	0	1	2
131	3	3	8	0	1	1
132	3	3	9	0	1	1
133	2	3	9	1	1	2
134	1	4	5	0	1	1
135	3	2	7	0	1	1
136	3	1	10	0	1	2
137	1	3	6	0	1	1
138	1	3	10	0	1	1
139	1	3	5	0	1	1
140	3	4	8	0	0	2
141	3	1	9	0	1	2
142	3	3	9	0	1	3
143	1	4	6	0	1	1
144	3	4	11	0	0	1
145	3	2	9	0	0	3
146	3	3	9	0	1	2
147	3	4	12	0	1	2

Appendix 6 (Continued)

OBS	ME	SYMPT	PB	HIST	BSE	DETC
148	3	1	11	0	1	1
149	1	3	10	0	1	1
150	1	3	9	0	1	1
151	3	2	6	0	1	1
152	2	3	5	0	1	1
153	1	3	7	1	1	1
154	1	3	9	0	1	1
155	1	4	6	0	1	1
156	2	3	8	1	1	1
157	3	1	8	0	1	1
158	3	4	5	0	1	1
159	3	3	8	0	1	1
160	3	4	8	0	1	2
161	2	1	10	0	1	2
162	1	4	9	0	1	1
163	3	4	10	0	1	1
164	2	4	9	0	1	1
165	1	4	6	0	1	1
166	3	1	7	0	1	2
167	3	3	5	0	0	1
168	2	3	5	0	1	1
169	2	4	9	0	1	2
170	3	3	9	0	1	1
171	3	1	5	0	1	1
172	2	1	5	0	1	1
173	3	4	6	0	1	1
174	3	3	6	0	1	1
175	1	4	6	0	1	1
176	3	3	10	0	0	1
177	3	4	6	0	1	1
178	1	3	5	1	1	2
179	3	2	9	0	0	1
180	1	3	7	1	1	1
181	3	3	10	0	0	1
182	1	3	8	0	1	2
183	2	2	7	0	1	2
184	3	2	9	0	1	1

Appendix 6 (Continued)

OBS	ME	SYMPT	PB	HIST	BSE	DETC
185	3	3	13	0	1	2
186	3	4	6	0	0	1
187	3	3	8	0	1	2
188	2	3	10	0	0	1
189	3	2	10	0	1	2
190	3	4	11	0	1	2
191	3	4	7	0	1	1
192	2	3	8	1	1	1
193	2	4	10	0	0	1
194	3	3	11	0	1	1
195	2	3	9	1	1	2
196	1	4	5	1	1	1
197	2	3	7	0	1	2
198	1	4	7	0	1	1
199	3	3	9	0	1	1
200	1	3	5	0	1	1
201	3	2	10	0	1	2
202	1	4	5	0	1	1
203	1	3	5	0	1	1
204	1	3	9	0	1	2
205	3	2	7	0	0	2
206	3	1	10	0	1	1
207	1	4	5	1	1	1
208	3	4	6	0	1	2
209	3	3	5	0	1	1
210	1	3	7	0	1	1
211	1	3	8	0	1	2
212	3	2	9	0	1	1
213	3	1	6	0	0	2
214	3	2	13	1	1	1
215	3	2	9	1	1	1
216	2	4	7	0	1	1
217	3	4	5	0	0	1
218	3	1	6	0	1	2
219	3	3	8	0	1	1
220	1	4	7	0	1	1
221	3	3	6	0	1	1

Appendix 6   (Continued)

OBS	ME	SYMPT	PB	HIST	BSE	DETC
222	3	3	5	0	1	1
223	1	4	5	0	1	1
224	3	3	6	0	1	1
225	3	4	9	0	1	2
226	1	3	8	0	1	1
227	3	3	5	0	1	1
228	2	2	8	0	1	1
229	1	4	6	1	1	1
230	2	4	6	0	1	2
231	3	2	8	0	1	2
232	2	1	10	0	1	1
233	3	2	7	0	1	2
234	1	4	5	0	0	1
235	1	3	6	1	1	1
236	3	2	11	0	1	2
237	3	4	5	0	1	1
238	3	4	8	0	1	1
239	3	2	7	0	1	2
240	1	2	7	0	1	1
241	3	3	11	1	0	3
242	3	1	5	0	1	1
243	2	4	5	0	1	1
244	1	4	6	0	1	1
245	3	1	8	0	0	1
246	1	4	5	0	0	1
247	2	3	6	1	1	1
248	2	4	8	0	1	1
249	3	3	6	0	1	1
250	3	1	11	0	1	3
251	3	3	9	0	1	1
252	3	3	10	0	1	1
253	3	1	5	0	1	2
254	1	3	5	0	1	1
255	1	4	5	0	1	1
256	2	4	5	1	1	1
257	3	4	6	0	1	1
258	3	4	7	0	1	1

Appendix 6 (Continued)

OBS	ME	SYMPT	PB	HIST	BSE	DETC
259	1	3	10	0	1	1
260	2	3	10	0	0	3
261	3	3	8	0	1	1
262	3	4	6	0	1	1
263	2	4	6	0	1	1
264	3	3	10	0	1	1
265	3	4	6	0	1	1
266	3	2	11	0	0	2
267	3	3	9	0	1	3
268	3	3	8	0	1	2
269	1	2	6	0	1	1
270	1	3	9	0	1	1
271	1	3	6	1	1	1
272	3	2	10	1	1	1
273	3	3	7	0	1	2
274	1	3	6	0	1	1
275	2	3	8	0	1	1
276	1	4	5	0	1	1
277	3	3	10	0	1	1
278	3	4	8	0	1	1
279	3	2	11	1	1	3
280	3	4	7	0	1	1
281	1	3	9	0	1	1
282	1	4	7	1	1	1
283	3	3	9	0	0	2
284	3	2	9	1	1	2
285	3	3	6	0	1	1
286	3	2	6	0	1	2
287	1	4	5	1	1	1
288	3	3	5	0	0	1
289	3	4	10	0	1	1
290	3	2	9	0	1	1
291	3	3	6	0	1	1
292	3	3	7	0	1	2
293	1	2	10	0	1	3
294	3	3	10	0	1	1
295	3	1	7	0	0	1

Appendix 6 (Continued)

OBS	ME	SYMPT	PB	HIST	BSE	DETC
296	2	4	7	0	1	1
297	3	4	6	0	1	1
298	3	3	12	0	1	1
299	3	3	8	0	0	1
300	3	2	8	0	1	2
301	3	4	6	0	1	2
302	3	3	10	0	0	1
303	2	3	5	0	1	1
304	3	2	10	0	1	2
305	1	4	8	0	0	1
306	1	4	5	0	1	1
307	3	4	11	0	1	2
308	2	2	10	0	1	1
309	2	3	5	0	1	1
310	2	3	5	0	1	1
311	3	3	5	0	1	1
312	1	4	6	0	1	1
313	3	3	7	0	1	2
314	3	3	7	0	1	3
315	3	2	6	1	0	2
316	1	4	5	0	1	1
317	2	4	5	0	1	1
318	1	4	7	1	1	2
319	3	3	10	0	1	1
320	1	4	8	0	1	1
321	1	4	8	1	1	1
322	3	3	9	0	1	1
323	3	2	11	0	0	3
324	1	4	5	0	1	1
325	3	3	10	0	1	1
326	1	4	6	0	1	1
327	1	4	5	0	1	1
328	3	3	10	0	1	2
329	1	3	7	0	1	1
330	1	4	6	0	1	1
331	3	2	10	0	1	1
332	3	2	7	0	1	1

Appendix 6   (Continued)

OBS	ME	SYMPT	PB	HIST	BSE	DETC
370	3	4	10	0	1	1
371	1	4	5	0	1	1
372	2	4	6	0	1	1
373	1	4	6	0	1	1
374	3	3	5	0	1	1
375	3	2	11	0	1	2
376	1	3	10	0	1	1
377	1	4	9	0	1	1
378	2	3	6	1	1	1
379	3	3	9	0	1	2
380	3	2	9	0	1	2
381	3	3	9	0	0	1
382	2	3	9	0	1	1
383	2	3	10	0	1	1
384	3	3	9	0	1	1
385	2	3	12	0	1	2
386	3	1	11	0	1	1
387	3	1	5	0	1	2
388	1	4	5	0	1	1
389	2	4	5	0	1	1
390	3	3	17	0	0	2
391	1	3	6	0	1	1
392	3	2	6	0	1	1
393	3	2	10	0	1	2
394	1	3	5	0	1	1
395	3	3	5	0	1	1
396	3	1	5	0	1	2
397	3	2	10	0	0	2
398	3	4	6	0	1	2
399	2	3	6	0	1	2
400	3	2	7	0	1	2
401	3	2	8	0	0	1

Appendix 6 (Continued)

OBS	ME	SYMPT	PB	HIST	BSE	DETC
402	1	3	9	0	1	1
403	1	4	5	0	1	1
404	3	2	12	0	1	2
405	3	3	10	0	1	2
406	2	3	5	0	1	1
407	3	3	11	0	1	2
408	1	3	10	0	1	2
409	3	4	8	0	1	1
410	1	4	6	0	1	1
411	2	2	6	0	1	1
412	3	4	7	0	1	1

References

Abbott, R. D. (1985). Logistic regression in survival analysis. *American Journal of Epidemiology*, **121**, 465–471.

Albert, A., and Anderson, J. A. (1984). On the existence of maximum likelihood estimates in logistic models. *Biometrika*, **71**, 1–10.

Begg, C. B., and Gray, R. (1984). Calculation of polychotomous logistic regression parameters using individualized regressions. *Biometrika*, **71**, 11–18.

Belsley, D. A., Kuh, E., and Welsch, R. E. (1980). *Regression Diagnostics: Identifying Influential Data and Sources of Collinearity*. Wiley, New York.

Bendel, R. B., and Afifi, A. A. (1977). Comparison of stopping rules in forward regression. *Journal of the American Statistical Association*, **72**, 46–53.

Bishop, Y. M. M., Feinberg, S. E., and Holland, P. (1975). *Discrete Multivariate Analysis: Theory and Practice*. MIT Press, Boston.

Box, G. E. P., and Tidwell, P. W. (1962). Transformation of the independent variables. *Technometrics*, **4**, 531–550.

Brenn, T., and Arnesen, E. (1985). Selecting risk factors: A comparison of discriminant analysis, logistic regression and Cox's regression model using data from the Tromsø heart study. *Statistics in Medicine*, **4**, 413–423.

Breslow, N. E., and Cain, K. C. (1988). Logistic regression for two-stage case-control data. *Biometrika*, **75**, 11–20.

Breslow, N. E., and Day, N. E. (1980). *Statistical Methods in Cancer Research*. Vol. 1 – The analysis of case-control studies. International Agency on Cancer, Lyon, France.

Breslow, N. E., and Zaho, L. P. (1988). Logistic regression for stratified case-control studies. *Biometrics*, **44**, 891–899.

Brown, C. C. (1982). On a goodness-of-fit test for the logistic model based on score statistics. *Communications in Statistics*, **11**, 1087–1105.

Bryson, M. C., and Johnson, M. E. (1981). The incidence of monotone likelihood in the Cox model. *Technometrics*, **23**, 381–384.

Cardell, N. S., and Steinberg, D. (1987). Estimating quantal choice models from pooled choice based samples and supplementary random samples without choice data. *Technical Report 87-1*. Dept. of Economics, San Diego State Univ., San Diego, CA.

Chambless, L. E., and Boyle, K. E. (1985). Maximum likelihood methods for complex sample data: Logistic regression and discrete proportional hazards models. *Communications in Statistics: Theory and Methods*, **14**, 1377–1392.

Claris Corporation (1988). MacDraw II, Version 1.1, Claris Corporation, Mountain View, CA.

Cook, R. D. (1977). Detection of influential observations in linear regression. *Technometrics*, **19**, 15–18.

Cook, R. D. (1979). Influential observations in linear regression. *Journal of the American Statistical Association*, **74**, 169–174.

Cook, R. D., and Weisberg, S. (1982). *Residuals and Influence in Regression.* Chapman Hall, New York.

Cooke, J. R., and Sobel, E.T. (1986). MathWriter™, Version 1.4, Mathematical Typesetting with the MacIntosh, Cooke Publications, Ithaca.

Copas, J. B. (1983). Plotting p against x. *Applied Statistics*, **32**, 25–31.

Cornfield, J. (1951). A method of estimating comparative rates from clinical data; Applications to cancer of the lung, breast and cervix. *Journal of the National Cancer Institute*, **11**, 1269–1275.

Cornfield, J. (1962). Joint dependence of the risk of coronary heart disease on serum cholesterol and systolic blood pressure: A discriminant function analysis. *Federation Proceedings*, **21**, 58–61.

Costanza, M. C., and Afifi, A. A. (1979). Comparison of stopping rules in forward stepwise discriminant analysis. *Journal of the American Statistical Association*, **74**, 777–785.

Cox, D. R. (1970). *The Analysis of Binary Data.* Methuen, London.

Cox, D. R. (1972). Regression models and life tables. *Journal of the Royal Statistical Society, Series B*, **34**, 187–220.

Cox, D. R., and Hinkley, D. V. (1974). *Theoretical Statistics*. Chapman Hall, London.

Cox, D. R., and Oakes, D. (1984). *Analysis of Survival Data*. Chapman Hall, London.

Crowder, M. J. (1978). Beta-binomial Anova for proportions. *Applied Statistics*, **27**, 34–37.

Day, N. E., and Byar, D. P. (1979). Testing hypotheses in case–control studies – equivalence of Mantel–Haenszel statistics and logit score tests. *Biometrics*, **35**, 623–630.

Dixon, W. J. (1987). *BMDP Statistical Software*. University of California Press, Berkeley.

Dobson, A. (1983). *An Introduction to Statistical Modelling*. Chapman Hall, New York.

Efron, B. (1975). The efficiency of logistic regression compared to normal discriminant function analysis. *Journal of the American Statistical Association*, **70**, 892–898.

Efron, B. (1986). How biased is the apparent error rate? *Journal of the American Statistical Association*, **81**, 461–470.

Efron, B. (1988). Logistic regression, survival analysis and the Kaplan-Meier curve. *Journal of the American Statistical Association*, **83**, 414–225.

Farewell, V. T. (1979). Some results on the estimation of logistic models based on retrospective data. *Biometrika*, **66**, 27–32.

Flack, V. F., Chang, P. C. (1987). Frequency of selecting noise variables in subset regression analysis: A simulation study. *American Statistician*, **41**, 84–86.

Fears, T. R., and Brown, C. C. (1986). Logistic regression methods for retrospective case-control studies using complex sampling procedures. *Biometrics*, **42**, 955–960.

Fleiss, J. (1979). Confidence intervals for the odds ratio in case–control studies: State of the art. *Journal of Chronic Diseases*, **32**, 69–77.

Fleiss, J. (1986). *The Design and Analysis of Clinical Experiments.* Wiley, New York.

Fowlkes, E. B. (1987). Some diagnostics for binary regression via smoothing. *Biometrika,* **74**, 503–515.

Freedman, D. A. (1983). A note on screening regression equations. *American Statistician,* **37**, 152-155.

Furnival, G. M., and Wilson, R. W. (1974). Regression by leaps and bounds. *Technometrics,* **16**, 499–511.

Gart, J. J., and Thomas, D. G. (1972). Numerical results on approximate confidence limits for the odds ratio. *Journal of the Royal Statistical Society, Series B,* **34**, 441–447.

Green, M. S. and Symmons, M. J. (1983). A comparison of the logistic risk function and the proportional hazards model in prospective epidemiologic studies. *Journal of Chronic Diseases,* **36**, 715–724.

Greenland, S. (1989). Modelling variable selection in epidemiologic analysis. To appear *American Journal of Public Health.*

Griffiths, W. E., and Pope, P. J. (1987). Small sample properties of probit models. *Journal of the American Statistical Association,* **82**, 929–937.

Grizzle, J., Starmer, F., and Koch, G. (1969). Analysis of categorical data by linear models. *Biometrics,* **25**, 489–504.

Guerro, V. M., and Johnson, R. A. (1982). Use of the Box-Cox transformation with binary response models. *Biometrika,* **69**, 309–314.

Halpern, M., Blackwelder, W. C., and Verter, J. I. (1971). Estimation of the multivariate logistic risk function: A comparison of the discriminant function and maximum likelihood approaches. *Journal of Chronic Disease,* **24**, 125–158.

Haseman, J. K., and Hogan, M. D. (1975). Selection of the experimental unit in teratology studies. *Teratology,* **12**, 165–172.

Haseman, J. J., and Kupper, L. L. (1979). Analysis of dichotomous response data from certain toxicological experiments. *Biometrics,* **35**, 281–293.

Hastie, T., and Tibshirani, R. (1986). Generalized additive models. *Statistical Science*, **3**, 297–318.

Hastie, T., and Tibshirani, R. (1987). Generalized additive models: Some applications. *Journal of the American Statistical Association*, **82**, 371–386.

Hauck, W. W., and Donner, A. (1977). Wald's Test as applied to hypotheses in logit analysis. *Journal of the American Statistical Association*, **72**, 851–853.

Hauck, W. W. (1985). A comparison of the logistic risk function and the proportional hazards model in prospective epidemiologic studies. *Journal of Chronic Diseases*, **38**, 125–126.

Hjort, N. L. (1988). Estimating the logistic regression equation when the model is incorrect. *Technical Report*, Norwegian Computing Center, Oslo, Norway.

Hosmer, D. W., Jovanovic, B., and Lemeshow, S. (1989). Best subsets logistic regression. Submitted for publication.

Hosmer, D. W., and Lemeshow, S. (1980). A goodness-of-fit test for the multiple logistic regression model. *Communications in Statistics*, **A10**, 1043–1069.

Hosmer, D. W., Lemeshow, S., and Klar, J. (1988). Goodness-of-fit testing for multiple logistic regression analysis when the estimated probabilities are small. *Biometrical Journal*, **30**, 911-924.

Hosmer, D. W., Wang, C. Y., Lin, I. C., and Lemeshow, S. (1978). A computer program for stepwise logistic regression using maximum likelihood. *Computer Programs in Biomedicine*, **8**, 121–134.

Hosmer, T., Hosmer, D. W., and Fisher, L. L. (1983). A comparison of the maximum likelihood and discriminant function estimators of the coefficients of the logistic regression model for mixed continuous and discrete variables. *Communications in Statistics*, **B12**, 577-593.

Jennings, D. E. (1986a). Judging inference adequacy in logistic regression. *Journal of the American Statistical Association*, **81**, 471–476.

Jennings, D. E. (1986b). Outliers and residual distributions in logistic regression. *Journal of the American Statistical Association*, **81**, 987–990.

Kalbfleisch, J. D., and Prentice, R. L. (1980). *The Statistical Analysis of Failure Time Data*. Wiley, New York.

Kay, R., and Little, S. (1986). Assessing the fit of the logistic model: A case study of children with haemalytic uraemic syndrome. *Applied Statistics*, **35**, 16–30.

Kay, R., and Little, S. (1987). Transformation of the explanatory variables in the logistic regression model for binary data. *Biometrika*, **74**, 495–501.

Kelsey, J. L., Thompson, W. D., and Evans, A. S. (1986). *Methods in Observational Epidemiology*. Oxford University Press, New York.

Kleinbaum, D. G., Kupper, L. L., and Morgenstern, H. (1982). *Epidemiologic Research: Principles and Quantitative Methods*. Van Nostrand Reinhold, New York.

Kleinbaum, D. G., Kupper, L. L., and Muller, K. E. (1987). *Applied Regression Analysis and Other Multivariable Methods*. PWS-Kent, Boston.

Lachenbruch, P. A. (1975). *Discriminant Analysis*. Hafner, New York.

Landwehr, J. M., Pregibon, D., and Shoemaker, A. C. (1984). Graphical methods for assessing logistic regression models. *Journal of the American Statistical Association*, **79**, 61-71.

Lawless, J. F., and Singhal, K. (1978). Efficient screening of non-normal regression models. *Biometrics*, **34**, 318–327.

Lawless, J. F., and Singhal, K. (1987a). ISMOD: An all subsets regression program for generalized linear models.I. Statistical and computational background. *Computer Methods and Programs in Biomedicine*, **24**, 117–124.

Lawless, J. F., and Singhal, K. (1987b). ISMOD: An all subsets regression program for generalized linear models II. Program guide and examples. *Computer Methods and Programs in Biomedicine*, **24**, 117–124.

Lee, E. T. (1980). *Statistical Methods for Survival Data Analysis*. Wadsworth, Belmont, CA.

Lemeshow, S., and Hosmer, D. W. (1982). The use of goodness-of-fit statistics in the development of logistic regression models. *American Journal of Epidemiology*, **115**, 92-106.

Lemeshow, S., and Hosmer, D. W. (1983). Estimation of odds ratios with categorical scaled covariates in multiple logistic regression analysis. *American Journal of Epidemiology*, **119**, 147–151.

Lemeshow, S., Teres, D., Avrunin, J. S., Pastides, H. (1988). Predicting the outcome of intensive care unit patients. *Journal of the American Statistical Association*, **83**, 348–356.

Lesaffre, E. (1986). Logistic discriminant analysis with applications in electrocardiography. Unpublished D.Sc. Thesis, University of Leuven, Belgium.

Liang, K. Y. (1987). Extended Mantel-Haenszel estimating procedure for multivariate logistic regressions. *Biometrics*, **43**, 289–300.

McCullagh, P., and Nelder, J. A. (1983). *Generalized Linear Models*. Chapman Hall, London.

Mallows, C. L. (1973). Some comments on Cp. *Technometrics*, **15**, 661–676.

Mauritsen, R. H. (1984). Logistic regression with random effects. Unpublished Ph.D. Thesis, Dept. of Biostatistics, University of Washington, Seattle.

Mickey, J., and Greenland, S. (1989). A study of the impact of confounder-selection criteria on effect estimation. *American Journal of Epidemiology*, **129**, 125–137.

Microsoft Corporation (1987). Microsoft® Word, Word Processing Program Version 3.1 for the Apple® Macintosh™, Microsoft Corporation, Bellvue.

Miettinen, O. S. (1976). Stratification by multivariate confounder score. *American Journal of Epidemiology*, **104**, 609–620.

Moolgavkar, S., Lustbader, E., and Venzon, D. J. (1985). Assessing the adequacy of the logistic regression model for matched case-control studies. *Statistics in Medicine*, **4**, 425–435.

Moore, D. S. (1971). A chi-square test with random cell boundaries. *Annals of Mathematical Statistics*, **42**, 147–156.

Moore, D. S., and Spruill, M. C. (1975). Unified large-sample theory of general chi-square statistics for tests of fit. *Annals of Statistics*, **3**, 599–516.

Numerical Algorithms Group (1987). The generalized linear interactive modelling system: The GLIM system. Release 3.77. Royal Statistical Society, London.

Pastides, H., Kelsey, J. L., Holford, T. R., and LiVolsi, V. A. (1985). The epidemiology of fibrocystic breast disease. *American Journal of Epidemiology*, **121**, 440–447.

Pastides, H., Kelsey, J. L., LiVolsi, V. A., Holford, T., Fischer, D., and Goldberg, I. (1983). Oral contraceptive use and fibrocystic breast disease with special reference to its histopathology. *Journal of the National Cancer Institute*, **71**, 5–9.

Peduzzi, P. N., Hardy, R. J., and Holford, T. R. (1980). A stepwise selection procedure for nonlinear regression models. *Biometrics*, **36**, 511–516.

Pierce, D. A., and Sands, B. R. (1975). Extra-bernoulli variation in binary data. *Technical Report No. 46.*, Department of Statistics, Oregon State University.

Pregibon, D. (1980). Goodness-of-link tests for generalized linear models. *Applied Statistics*, **29**, 15–24.

Pregibon, D. (1981). Logistic regression diagnostics. *Annals of Statistics*, **9**, 705–724.

Pregibon, D. (1984). Data analytic methods for matched case-control studies. *Biometrics*, **40**, 639–651.

Prentice, R. L. (1976). A generalization of the probit and logit methods for dose response curves. *Biometrics*, **32**, 761–768.

Prentice, R. L. (1986). A case-cohort design for epidemiologic cohort studies and disease prevention trials. *Biometrika*, **73**, 1–11.

Prentice, R. L., and Pyke, R. (1979). Logistic disease incidence models and case-control studies. *Biometrika*, **66**, 403–411.

Rao, C. R. (1973). *Linear Statistical Inference and Its Application*, Second Edition. Wiley, New York.

Rothman, K. J. (1986). *Modern Epidemiology*. Little Brown, Boston.

Santner, T. J., and Duffy, D. E. (1986). A note on A. Albert's and J. A. Anderson's conditions for the existence of maximum likelihood estimates in logistic regression models. *Biometrika*, **73**, 755–758.

SAS Institute Inc. (1988). *SAS Guide for Personal Computers, Version 6.03.* SAS Institute Inc., Cary, NC.

Schaefer, R. L. (1986). Alternative estimators in logistic regression when the data are collinear. *Journal of Statistical Computation and Simulation*, **25**, 75–91.

Schlesselman, J. J. (1982). *Case–Control Studies.* Oxford University Press, New York.

Scott, A. J., and Wild, C. J. (1986). Fitting models under case-control or choice based sampling. *Journal of the Royal Statistical Association, Series B*, **48**, 170–182.

Statistics and Epidemiology Research Corporation (1988). *EGRET Statistical Software.* SERC Inc., Seattle, WA.

Steinberg, D., and Cardell, N. S. (1987). Estimating logistic regression models when the dependent variable has no variance. *Technical Report 87-28*. Dept. of Economics, San Diego State University, San Diego, CA.

Truett, J., Cornfield, J., and Kannel, W. (1967). A multivariate analysis of the risk of coronary heart disease in Framingham. *Journal of Chronic Diseases*, **20**, 511–524.

Tsiatis, A. A. (1980). A note on a goodness-of-fit test for the logistic regression model. *Biometrika*, **67**, 250–251.

Velleman, P. F., and Velleman, A. Y. (1988). *Data Desk Professional: Statistics and Reference Guides.* Odesta Corp., Northbrook, IL.

White, H. (1982). Maximum likelihood estimation of misspecified models. *Econometrika*, **50**, 1–25.

White, H. (1989). *Estimation, Inference and Specification Analysis.* Cambridge University Press, New York.

Wilkinson, L. (1987). *SYSTAT: The System for Statistics.* Systat Inc., Evanston, IL.

Williams, D. (1975). The analysis of binary responses from toxicological experiments involving reproduction and teratogenicity. *Biometrics*, **31**, 949–952.

Williams, D. A. (1982). Extra-binomial variation in linear logistic models. *Applied Statistics*, **31**, 144–148.

Wu, M., and Ware, J. H. (1979). On the use of repeated measurements in regression analysis with dichotomous responses. *Biometrics*, **35**, 513–521.

Index